Humming

THE STUDY OF SOUND

Editor: Michael Bull

Each book in The Study of Sound offers a concise look at a single concept within the field of sound studies. With an emphasis on the interdisciplinary nature of the topics at hand, the series explores a range of core issues, debates, and objects within sound studies from a variety of perspectives and within a multitude of contexts.

Published Titles:

The Sound of Nonsense by Richard Elliott

Forthcoming Titles:

Sirens by Michael Bull

Sonic Intimacy by Malcolm James

Sonic Fiction by Holger Schulze

Humming

Suk-Jun Kim

BLOOMSBURY ACADEMIC
NEW YORK • LONDON • OXFORD • NEW DELHI • SYDNEY

BLOOMSBURY ACADEMIC
Bloomsbury Publishing Inc
1385 Broadway, New York, NY 10018, USA
50 Bedford Square, London, WC1B 3DP, UK

BLOOMSBURY, BLOOMSBURY ACADEMIC and the Diana logo are trademarks of Bloomsbury Publishing Plc

First published in the United States of America 2019

Cover design and image by Liron Gilenberg www.ironicitalics.com

A catalog record for this book is available from the Library of Congress.

ISBN: HB: 978-1-5013-2460-4
PB: 978-1-5013-2461-1
ePDF: 978-1-5013-2459-8
eBook: 978-1-5013-2462-8

Series: The Study of Sound

Typeset by RefineCatch Limited, Bungay, Suffolk
Printed and bound in the United States of America

To my wife, Sungeun

CONTENTS

PREFACE: TACTILITY OF HUMMING

One cannot find an opportunity of knowing things if they are shown and experienced only through their smooth surface. The polished, even surface is, in a sense, an abstraction of things, a closed form, which would, perhaps, be sufficient if our aim were only to appreciate things as they are. Knowing things, however, expects us to move further than that. It is an act of seeking a gap, a crack, or an incongruity and making it visible, tactile on that calm and placid veneer. It is through this act that the surface of things becomes an open form.

Humming, to most of us, is a closed form. It is as such, not through a hard, glossy surface, but through its very abstractness. It is evasive and encourages us to look away, and yet, it captivates us, wrapping us in its spell. It is thus very hard for us with a mind of modern science – that is, a mind with the tendency of being objective about things – to know humming in its fundamentally aerial form. For me, humming became a subject for knowing when I discovered a gap, which, when I first identified, felt like an inconsequential and insignificant feature; that humming means two things: hums that we can make with our mouth shut, either tunes that we can hum or the very act of humming; and hums that have nothing to do with us, ones that we do not want to make, ones that we do not want to hear, such as electronic hums or traffic hums. What does this gap mean, this distance, and why such a distance? I had to ask. And this question led to more questions for which I wanted to find an answer, and through the initially unbridgeable distance emerged a certain field to posit and stand on. And just like that, this book started.

In the first chapter, '*My hums? . . . Just about hums?*', I question this very gap that seems to have curiously been neglected by the literature of sound studies. As I bring it to the fore and confront it, I take particular issue with the fact that the emergence of humming is in itself a secret, that is, the very production of humming is through one's mouth shut and lips sealed, an act of secrecy. This places humming in an unusual topology in relation to the study of voice, and specifically, that of psychoanalysis in which the voice, along with the gaze, is considered a key contribution of Lacan. In this chapter, therefore, I ask questions on the secrecy of humming through the lens of psychoanalysis, particularly, the discussions made by Dolar, Miller and Žižek, hoping to identify key features of humming and possible methodologies with which to examine its secrecy. In the second chapter, *The Secrecy of Humming*, I examine the peculiar relation of humming to its origin, the mouth, and how silence is borne out of this relation. I do this by conducting a case study of John Cage's infamous performance/lecture piece, *Lecture on Nothing*. Much has been examined on *Lecture on Nothing*, but in relation to humming and its ontology of negation, I take a slight, yet, decisively different view on the piece, highlight four modalities of Cage's silences. In doing so, I move onto the hums of the other. In the third, and final chapter, *Hums of the Other*, I explore the otherness of hums, first, by way of stories by Calvino – *A King Listens*, *Invisible Cities*, and *Six Memos for the Next Millennium* – and by Kafka – *The Castle*, *Metamorphosis*, and *The Burrow.* This is followed by the examination of three features of humming – the aphonic, the acousmatic, and the air – which leads our discussion of humming back to the mouth and its oral imaginaries. Finally, I take humming as symptoms understood by Lacan's – and Žižek's – psychoanalysis.

While many of the ideas in *Humming* had been developed through various artistic projects in the past, I could not have examined them as coherently and expansively as shown here without the help of Michael Bull, who saw potential through the quirkiness of the topic I proposed. I am also grateful to my editors – Susan Krogulski, Leah Babb-Rosenfeld, and Katherine De Chant (I happened to have three editors!) – at Bloomsbury Academic who graciously waited for and worked through my final draft. My thanks should also go to my colleagues at the Department of Music, University of Aberdeen, who have given me such understanding and confidence over the years and with whom I thoroughly enjoy working, to Gary Kendall who let me know that writing is like composing a piece of music, and to Paul Koonce, my mentor and friend, who taught me how to think.

1

'My Hums? . . . Just About Hums?'

'Do you have songs that would remind you of your childhood?'
'Yes.'
'Could you hum some for me?'
'My hums? . . . Just about hums?'

This short book on humming began almost ten years ago. In 2009, I was in Berlin, working on a series of sound projects. One such project was to collect the hums of different people from Berlin. Titled *In Tune, Out of Tune*,[1] the project required these different people to hum a tune that would remind them of their childhood. Collecting hums from approximately fifty to sixty people in Berlin to realize this project was a fascinating, and in a way, uneasy experience. Asking people who are strangers to you to hum a tune creates a series of fleeting moments that are tense and awkward. These moments, however short-lived and insignificant initially, are deeply felt; both by the person asking for the hum, and the person humming. The person doing the humming is left with the realisation that this task is considerably more difficult than he or she thought it to be.

Inviting and estranging

It was not at first my desire to observe the behaviour of those who agreed to offer hums. But those awkward moments never failed to stop, intervene, and cut through the normality of whatever social, emotional engagements the person and I had had just before. Being strangers to each other, the social construct agreed upon between us was flimsy and suspected, bringing us to see and hear what humming did to those who hummed and those who listened to it. You can witness this action, this process of humming, for yourself. Ask anyone close to you, either in terms of physical or social distance; a stranger would certainly be better for this experiment of humming a tune. In order to ask someone to hum for you, you first need to consider how to approach him or her. This is not an easy task, precisely because you are well aware of what this request signifies: you are seeking the person's permission to be invited into his or her intimate, personal space. In a sense, it is as if you have been invited to their house, all of a sudden and without much chit-chat. If they agree to hum, you are in a contract with them bound by a certain trust, one that usually would take longer and require considerably more effort to build. But to your surprise, you realize that *you already are in it*. And soon, they know that they are in it, too, and often, they realize that such trust is not what they had agreed on.

How would they start humming? They do not; they hesitate – they breathe in, and out, make a few coughs, try to clear their throat and tell a couple of stories about the tune and why this tune and not others. They smile, start the hum, but then stop it immediately and apologize for their clumsiness, either because they may have started it with a wrong note, either too high or too low in the register, or because they may have not been able to control tension in their throat, thereby producing an abnormal or unexpected sound – 'that's not my tune!' they may say. They may have completely and suddenly lost the tune, mumbling some of the lyrics as if to re-learn it, trying to bring it back to the melody they know. They wonder how the song could become so strange and forgotten to them: they roll their eyes high up into their skull, searching for the fleeting tune. They swing their body to and fro, sideways, stretching the chest, biting their lips, swallowing the silence, their breath heavy. Then they start humming again.

Once it starts, however, the listener and the 'hummer' are soon in the humming. It particularly helps if you know the tune, but it really doesn't matter much. Smiling is common in many occasions, and sometimes you can notice certain feelings – cheerfulness and gaiety, sorrow and sadness, fondness and welcoming, or melancholia and a sense of loss – that may not otherwise be identified and shared. Furthermore, humming can reveal rather ghostly and mysterious incidences that are usually hidden deep in our consciousness. I remember in two occasions when I became keenly aware of this. One was while I was collecting hums from Gloria Maya in 2012, who was then a professor of printmaking at Western New Mexico University in Silver City, New Mexico, USA. I was working on a humming project called *Silver City Humming*, and she decided to offer her hum of a children's song together with her sister, Nellie. Since Nellie was in another town, she called her up and they hummed together over the telephone.[2] There are three most fascinating – and perhaps, you could say, eerie – aspects about their hum that struck me while listening to them humming on that first occasion and haunted me again and again when I played it back in the studio. Being sisters, the voice of Gloria – whose body was present with me – is uncannily similar to that of Nellie, whom I had never met before. Humming, not singing with lyrics, enhanced the likeness between the sisters, resulting in a strange image of Gloria humming along with a recording of her own voice – her double, her simulacrum. Just before they start humming, they make sure that they are both ready to do so:

Nellie: *Okay. Are you ready, Gloria?*
Gloria: *Okay, okay. Are you ready?*
Nellie: *Yes, Okay.*
Gloria: *Okay.*

During the short amount of time the two sisters 'okay' with each other, I lose track of who is okaying who. Bizarrely, Gloria's simulacrum is on an equal standing, gaining the same power she has. They start humming well together at first, but gradually, Gloria appears to forget the tune and slowly loses synchronisation with Nellie. Perhaps this was because Gloria became aware that Nellie would have had no way of knowing what was going on between me and her sister in this far away room, or maybe it was just that

Gloria could not keep up with Nellie. No matter what the cause was, the effect is striking. At some point, Gloria stops humming completely and we have only the humming of Nellie through the crackling phone line. Almost at the same time that Gloria stops humming, Nellie's humming becomes stuck in one phrase of the tune, failing to find the resolving note. She continually returns to the note of the dominant chord, thus repeating the same phrase again and again, as if her humming had been on the looped groove of an LP record. The result is as if Gloria and her present body were subsumed by Nellie, her ghostly double, but then almost immediately, that ghost were also subsumed by the very power of subsuming – repetition and doubling, two operations that result in Freud's uncanny (*das Unheimliche*).[3] Gloria and I had to let Nellie hum for another ten minutes or so as she kept going and we did not want to embarrass her by stopping her. Gloria explained to me about the song before they hummed. I think she said that the tune was a common birthday song, popular among Chicanos (Americans of Mexican descent). And then she mentioned that Nellie's husband had suddenly passed away quite recently. Being a South Korean and not being familiar with the Chicano culture, the tune was completely foreign to me, even with the lyrics. And yet, I could not help but recognize the irony caused by the conflict between a sense of loss on the one hand, and that of hope on the other, which I could read neither from Gloria's face nor that of Nellie, but from their heartfelt humming.

On another occasion, I was visiting a nursing home in Aberdeen, Scotland, for another humming project titled *Aberdeen Humming*.[4] There was an old lady who agreed to offer her hum, but she kept stopping and apologizing as she felt she could not hum as well as she thought she could due to her recent surgery on her throat to remove tumours. Another lady in a wheelchair, who seemed to have been troubled by memory loss, said to me, repeatedly, her voice wasn't very good, and her humming wasn't very good, either. In both incidents, humming bore witness to and was symptomatic of the physical or mental difficulties that the women had been suffering from. At the same time, it brought me into their suffering and pain more immediately, directly, and gently than in any other way.

Humming is inviting. In an instant, humming puts us into a socio-acoustic cocoon as it erects a wall of intimacy and emotion. Its effect is at first towards oneself who hums as its resonance soothes his or her body and its tune brings back memories. In this sense, humming invites oneself to oneself; it guides and directs oneself to a path to knowing oneself whose constitution is sonic, and thus, to the extent of which, humming is epistemological. It is a grasping of the resonant self, whose body and memories – as well as the acoustic threshold of otherness – all contribute to a feedback system that probes into the surface of the body: into the throat, into the larynx, and into the lungs, reversing itself by pushing air out from the lungs to the larynx, throat, mouth and nose, expelling itself back into the world. When offered to others as an invitation – usually to those who are close to themselves, and in rare occasions, to strangers – the emotive and intimate attributes of humming do not lose their power but in fact expand and are empowered by sharing. That humming acts to conceal the tune's lyrics (as most tunes have them) is a characteristic of its power, for by removing the textural appearance of the tune (because you cannot completely remove its textural structure), it opens up possibilities for a true faculty of the voice. The sound of vocalic uttering, unhindered by the structure of language, demonstrates how the voice is produced, connecting to the apparatus of the mouth, head, shoulders, and upper body, as well as the lungs, through which air is expelled. This concealment is what constitutes humming, a topic to which I shall return later. For now, it is important to note that this concealment of the lyrics while humming is neither a complete nor a successful task. For humming, when offered to oneself or to others, does not aim to hide the lyrics, even though it requires the hummer to do so. In fact, humming functions regardless of whether or not the text of the tune is recognized. In the most distant awareness during humming, we may not recognize the tune or its lyrics, if it is a song, but the fact we can still hear someone humming is enough for a socio-acoustic relationship to be established. Furthermore, if we recognize the tune ('hey, I know the song!'), we are invited to participate in it, and thus, listening to humming can become a collaborative act if we choose it to be. Or perhaps, we cannot choose it to be, but we have already ended up participating in it as soon as we realize that we know the tune. Finally, when we know the lyrics, humming then takes another turn that binds us completely to its spell; a linguistic communication is made without linguistic means. The effect of humming that binds us to its act – making us hear, listen, and eventually participate in it – takes only an instant. How does humming, despite being a gentle and unassuming act, do its magic at such a speed? What causes its velocity? One answer, although it sounds antinomic, may be because humming alienates us.

Yes, humming is estranging, and thus, is estranged from the self. While pulling us into it, it, at once, pushes us out. Humming makes us worked up and makes us acutely aware of the failing features of ourselves, of the frail link between the voice and the body: the voice that isn't good, the breathing that is not enough, the throat that aches and cracks, the melody that goes round and round, or the memory that becomes weak and faint. Interestingly, the fact that humming makes us aware of this alienation, this disjointedness of the voice from the self, is clear evidence that it is not (just) the language that causes the voice to be estranged from the self, but something fundamental about the voice that is foreign to it. Mladen Dolar highlights that foreignness of the voice which assumes the link between the two – language and body – but never actually belongs to either:

> The voice ties language to the body, but the nature of this tie is paradoxical: *the voice does not belong to either.* It is not part of linguistics, which follows from my initial argument (after all, Saussure himself spoke of the non-phonic nature of the signifier; Derrida will insist on this at great length in Grammatology), but it is not part of the body either – not only does it detach itself from the body and leave it behind, it does not fit the body either, it cannot be situated in it, 'disacousmatized.' . . . So the voice stands at a paradoxical and ambiguous topological spot, at the intersection of language and the body, but this intersection belongs to neither. *What language and the body have in common is the voice, but the voice is part neither of language nor of the body.*[5]

Dolar further discusses the paradoxical nature of the relation of the 'non-voice' to language elsewhere in his book, where he suggests that the ontology – the *topological spot* – of the non-voice, which, he lists, are 'from coughing and hiccups to babbling, screaming, laughing, and singing', is never independent from linguistics, as if,

> by their very absence of articulation (or surplus-articulation in the case of singing), they were particularly apt to embody the

> structure as such, the structure at its minimal; or meaning as such, beyond the discernible meaning.[6]

Dolar continues to argue that these non-voices assume the voice's 'zero-point: the voice aiming at meaning,' which is to say, these non-voices usually mean something, such as a waiter's 'ahem' which starts Connor's book *Beyond Words*, where Connor argues it is and should be understood as meaning;[7] and as such, these non-voices are never external to linguistics. Although he did not specifically include humming in his list of non-voices, it is no doubt that, being a non-phoneme that is devoid of articulation, it is a non-voice. It is important to note that despite what Dolar argues, not all of the voice aims at meaning, if he means by 'the voice' at this point, those which include the voice proper as well as these non-voices; we can argue that humming, for example, does not really aim at meaning. Instead, it is after presence, a presence of being, of a state, of a feeling, and meaning seems to be attached later. You could also make a similar argument for other non-voices, such as coughing, hiccups, or (some but not all kinds of) laughing, which are involuntary. Without intention, these non-voices do not aim for anything but a presence. What is special about humming, it seems, is that regardless of a voluntary, intentional voice, humming doesn't mean anything in particular. I believe this has made humming elusive to put in any particular discourse: it is difficult to locate humming in purely vocalic practice, thus under the lens of the body and sound studies, because it is right before being mute and does barely engage with the body; and it is also fairly nonsensical, making it useless in linguistics, in communication; it's just a hum. Where do we place it then? Where is the topological spot of humming, if, as Dolar insists with the voice, it belongs neither to the body nor to language? Dolar seems to argue that it is *in the air*:

> [The voice] floats, and the floating voice is a much more immediately striking phenomenon than the floating signifier, *le significant flottant*, which has caused so much ink to flow. It is a bodily missile which has detached itself from its source, emancipated itself, yet remains corporeal.[8]

We will revisit the relationship between humming and the air later in the final chapter of this book, but at this point, it may suffice to say that there are, I argue, no other vocalic sounds – except, perhaps, laughter[9] – that are lighter, and thus, that float more freely than humming. Humming is directly connected to the air; the air is the culprit of its existence. Humming a tune consumes more air than others and immediately reaches to the throat and the lungs, the organ of the air. And it should also be pointed out that being or becoming lighter, or floating in the air does not necessarily mean that it – the voice, and for us, humming – will always move up, towards the sky; rather, floating, in this case, signifies a detachment from surfaces, which are propped from every direction, a levitation from the force of pulling, an estranging of itself from those who hum it.

In this manner of floating and estranging, humming can distance us from mundanity, its familiarity that would otherwise put us at ease and receptivity. It makes us sensitive to our hearing that comes to us as flawed, or the intention that the humming is suspicious or feared to be malicious (for example, a serial killer's humming). In the most muted form of the mouth's vocalic production, humming can nevertheless be thought of as a striking vocalic sound that would exemplify the strangeness or absurdity of an event or situation as any other vocal utterances like laughter or crying. The brief humming of the then Prime Minister David Cameron on 11 July 2016, immediately after finishing the announcement of his resignation from his office as the consequence of his defeated campaign against Brexit, which was captured and subsequently mocked by media around the world,[10] is a great example of the mischievous (or even suspicious, and for some, spiteful) humming. Obviously, the Prime Minister did not realize that the wireless microphone was still on when he turned back from the TV cameras, but this seemingly cheerful humming of his was at great odds with the gravity of the announcement he had just made. While his hum was just for a couple of seconds, it was somehow more than enough to highlight the incongruities and absurdity that the result of the historic referendum had caused.

Just, or more than, hums

Therefore, a hum is no longer, or has never been, *just* a hum. Through its pushing and pulling, humming engages us in some of the most unexpected, fascinating ways. Still, or perhaps, for the very reason that a hum is more than just or nothing but, the innocuous response of an old lady – *my hums? . . . just about hums?* – when asked to hum a tune or two for me is telling in many ways and needs further examination. First of all, I have no doubt that by the 'just', she did mean 'only' or 'mere' because humming is indeed a mere, simple, act, nothing but extraordinary, and at the same time, it may have been an expression of her disappointment; after all, she could have offered *more than hums*. She is saying – *there is and should be more to the songs than just hums, I can sing, because it's got some nice lyrics; there is a story, my story, my personal and intimate story that I can tell you, ah, memories, warm feelings, fun, joy, sadness, sorrow, melancholia . . . there is just too much, an abundance of things to it, but are you only interested in my hums, just about hums?* And thus, the 'just' is also a sense of frustration, disturbed by the confrontation of *abundance* – what is more than it needs to be – and *absence* – what is just, mere, nothing but. Furthermore, this frustration is conditioned by the fact that in order to hum, you would need to shut your mouth, effectively blocking the flow of that abundance of things that would have, should have, or could have been shared, and therefore, the sense of frustration experienced when asked to, agreed to, starting to hum is that of feeling a clot, a lump, in the chest, in the throat, and in the mouth, the last of which has been particularly tightly fastened. The fastening of the mouth, the only real term demanded by humming, leaves those which are in the mouth – the tongue, the teeth, and the saliva among others – which would normally play a crucial role in the functioning of the mouth, unwanted, redundant, and out of work. The *surplus* of the apparatus of the mouth, resulting from the *lack* of agency imposed by humming, is a prominent feature of hums. What are you going to do with your tongue when you hum? Where does it, or should it go? What will you do with the flowing saliva that gradually fills up in your mouth, threatening to choke your airway?

We will get to the mouth by way of humming in greater detail later in the next chapter. For now, I would like to posit one aspect of humming, a potent feature that contributes so much to our fascination with the act, to which I do not know how to have access, except by way, first, of the discussion of Lacan's 'mirror stage'. I will follow with a discussion of the development of the gaze and the object voice as psychoanalytic considerations, leading to an examination of humming's peculiar position within the object voice.

Miller notes the disparity in the development of Lacan's two additions of the objects (refer to Dolar's chapter 'The Object Voice'[11] for more information about Lacan's object voice) to the already established Freudian objects, the oral object and the anal object.[12] These are the scopic object – the gaze – and the object voice – the voice – where the former was advanced through a rather 'chance' encounter with his examination of the mirror stage about the same time as the publication of Merleau-Ponty's *The Visible and the Invisible*. While observing the role of a mirror, which materializes the imaginary identification of 'I see myself seeing myself,' Lacan saw the mirror hiding the crucial difference between vision, which is 'a function of the organ of sight,' and gaze, '[vision's] immanent object, where the subject's desire is inscribed.' Distinct from vision in that it is 'neither an organ nor a function of any biology,' the gaze became aptly identified as an object of desire, a desire for scoping. The latter, the voice, did not have such a clear journey of development by Lacan, but, Miller argues, a similar development trajectory can be made based 'on the model of the articulation between eye and gaze – without even needing a meditation such as that of the mirror.' Miller continues:

> The mirror is necessary to produce the 'seeing oneself', whereas the 'hearing oneself' is already present at the most intimate point of subjectivity – or to express it like Husserl, in 'the self-presence of the living present of subjectivity' [la présence à soi du présent vivant de la subjectivité].[13]

Following Miller, Lacan's use of the mirror in the development of the gaze as a scopic object, and the disclosing of the enmeshed workings of seeing, enacted by those between vision and gaze, is perfectly graspable. What is less convincing for me, though, is not the lack of a similar device in the development of the object voice, but rather, that Miller does not go into a detailed examination of the reason for such a lack. I think a more satisfactory explanation could be offered by examining how 'hearing oneself' can *already* be 'present at the most intimate point of subjectivity'. This may be explored by distinguishing the direction at which these two different objects – the gaze and the voice – aim. By design – both as a physiological construction and as a psychoanalytical object – the gaze always aims outwards. Even when the object of its aim is the self, the body, it is always already presented to the gaze as an external object. Edward S. Casey, in his book *The World at a Glance*, evinces the exclusive, interpretive, and introspective nature of the gaze that, through its enactment, repudiates the body and the self. Casey posits that when the gaze directs us to scrutinize and scan the world, our body ends up being the last thing that we take into account. And thus, he continues:

> Just as I distance myself from the other in the Sartrian *regard*, so I distance myself from myself in the course of the gaze. In the one case, alienation from the other (and vice versa) results; in the other, alienation of myself from myself: self-alienation in the form of disembodiment.[14]

The gaze cannot look back on the self as there is no way of gazing in; it is always gazing *out, pointing and darting out*. For the gaze, thus, to face the gazing-self would require the self either to become asomatous – an extracorporeal, disembodied or out-of-body state, or an incorporeal state – which, if ever possible, would be less a gaze than voyeurism, or to make use of an external object that could somehow reflect the gaze back to the self, such as a mirror. Still, even with the mirror, it is an attempt that is different from looking back on the self, for as Miller argues above, 'the mirror is there to materialize the image', not to materialize the gaze, let alone the self. But the image of what? Miller says it is of *I see myself seeing myself*, the image of *ocular feedback*. What is peculiar about the gaze and the mirror, though, is that it is not just the construction of the mirror through which this ocular feedback operates, but it is also the gaze itself that initiates and sustains that operation. Casey goes on to point out that there are two features of the gaze that have made it attractive for religious thinkers and scientists, *taking in* and *getting at the depth*, and explains further how these are exercised by the gaze:

> A gaze aspires to take in – not just to apprehend but to internalize – what it is engaged with. Its concern is not with noticing as such, much less with altering its content in any way. To gaze at something is to immerse oneself in it to such a degree that this

> something becomes part of one's ongoing perceptual history: a resource for one's subsequent looking. At the same time, gazing is getting into the depth of the gazed-at. If it takes something into oneself in the interiorization just mentioned, it also takes itself into the interior of the same something. Thus it combines two modes of interiority, that within (or coincident with) oneself and that without (identical with the depths of the content or scene).[15]

The gaze always strives to aim at depth; in other words, it cannot aim for the surface, which is in this occasion, the mirror. Directed at the mirror, therefore, the gaze cannot help but to face the *image* of the gazing self reflected onto – *inside, in the depth of* – the mirror, and again to its further depth, which is the image of the self who is gazing at the image of the self, and so on. In this way, the gaze is engaged in the perpetual introspection.

On the other hand, while there is a similar, feedback-like operation at work with the voice, it is neither perpetual nor purely introspective. First of all, the voice does not need an external object like a mirror for the voicing-self to hear the self; the body of the self plays the role of the mirror[16] as it resonates with the voice. While the gaze shoots outwards, aims at the depth of things, the voice aims at all, both the surface and the depth, as it permeates through all directions, in and out; even with the intention of the self aiming his or her voice at one direction, *psst!*, the self-body reflects it back to the self; furthermore, *hushed* or *psst-ed* voices are always suspect for being heard by others anyway. In this way, the object voice carries with it the body of the self. Second, unlike the insatiable drive of the gaze, the voice does not turn into perpetuation. There exists no thread of 'I hear myself hearing myself . . . and so on'; you utter something with your mouth, and you hear yourself making that sound, and that is it. And perhaps, there lies another fundamental difference between the gaze facing the mirror and the voice resonating through the self's body: while the gaze appears to be invested in its own, linear feedback system and not be moving away from it while in operation, the voice appears to be invested in two distinctive processes, 'hearing oneself' and 'listening to oneself', which separate and sever the feedback once the voice-hearing self crosses them. Miller points out one paradox involving the 'subject's perception of his own speech,' discussed by Lacan, where the subject always needs to hear oneself in order to speak:

> [H]is own speech contains a spontaneous reflexivity, so to speak, a self-affectation that never fails to charm the analyst of consciousness phenomena. But this 'hearing oneself' is different from 'listening to oneself', where the subject's applied attention corrects, comes to recapture this spontaneous reflexivity. On this point, we note that the subject cannot listen to himself without being divided; numerous experiments show that, for instance, if the subject gets his own speech fed back to him with a small delay through headphones, he will lose the thread of his speech if he listens to what he says.[17]

There is a hard line between this division of the self, between the one who hears his or her own voice and the other who listens to it, and surprisingly, they are exclusive with each other; you cannot be both hearing and listening to yourself at the same time. For instance, most vocalic sounds, including speaking or singing, require that you listen to yourself. With others like coughing, laughing, sneezing, or other non-voices, you cannot do anything but hear yourself making them (it would be fascinating to be able to listen to yourself coughing, or scary to listen to yourself laughing).

Except humming, that is. This is because humming is liminal. With humming, interestingly, the line is blurred; you may be hearing *or/and* listening to yourself humming, and you could continue to be on the boundary between hearing and listening while humming. The scission between ear and voice, and its antinomic tension, much examined and highlighted by Miller who advances Lacan's split between eye and gaze, seems, even for a short time, to be obscure and pacified. In other words, from the perspectives of humming, the self is either never divided, or has already been divided, but it is never *being divided* by the voice; that pain of separation does not occur. (Note that what happens when you are asked to hum to a stranger or in an uncommon or unfamiliar situation is, in a sense, a sudden jolt of realisation of that divide.) What would contribute to such a feat of humming to be on both sides? There are a few attributes of humming that make this possible. First, humming, when directed at the self and only the self, is noncommittal and demands little effort, energy, or emotional investment. And it is a mundane and insignificant, and at the same time, intimate and personal act. Despite being quotidian, however, humming is an intentional act with a certain aim, requiring a certain degree of control, although the intention is equivocal, the aim is vague, and the control tends to fail easily. Furthermore, hiding the lyrics while humming would make it easier to perform the dual act of hearing and listening as there is not much pressure to correct oneself (although they need to watch out for wrong notes). But most of all, humming is special among the vocalic sounds in that it is specifically directed inward, more than any other sounds. The natural omnidirectionality of the voice is hampered by the mouth shut, lips sealed, which reflect humming back to the self and amplify the function of hearing, *listening in*. In this way, humming always swirls toward the interior *initially*. And again, in this regard, the closed mouth and its sealed lips are functioning like a mirror; but unlike a mirror, they are not there to materialize the aural image, but to drive humming toward the interior. If the gaze is the scoping object, perhaps the hum is a hearing-listening object, or object *h*, to borrow the famous Lacanian signification, where *h* could point to the dual – coupled, ambiguous, and even deceitful and treacherous – nature of the hum. Steven Connor recognizes the dubious nature of *h* in phonology where even if it is usually considered to be a kind of consonant, it does not function in the way normal consonants are thought to function, such as 'stopping, detaining or detouring the efflux of air through the mouth.' He further highlights the eccentric characters of *h*:

> The sound supposed to be signalled by the aspirate letter *h* is a kind of pure debouchure, orally unobstructed and minimally modified by the mouth. It approaches, we might say, the degree zero of consonance. It is a consonant in the sense that it is lacking in voice, but vowel-like in that it appears open and unconstrained. Indeed, it has sometimes been described as an 'aspirated vowel'. Aristotle notes that 'we cannot use the voice when breathing in or out, but only when holding the breath; for one who holds his breath produces the motion by its means.' It is the absence or minimal presence of this checking in the case of *h* which means that it belongs wholly neither to voice nor to noise, neither to larynx nor to mouth.[18]

Furthermore, it is not just *h* that lends to the ambivalent characteristics of the hum. A hum is a construction of *h* with '*um*'; a holding of the *h*, the air, the questionable 'aspirational vowel', the unconstrained and intractable (non-)sound. But we are getting ahead of ourselves.[19] For now, though, let's go back to the drive of the hum that directs us inward. As I said, the closed mouth with sealed lips reflects the humming back to the self and towards the interior. But this interior is not what we would normally categorize as such, such as the mind, the soul, the spiritual, or whatever. It is a void, a space of nothingness where the self encounters the Other:

> If the voice as object *a* does not in the least belong to the sonorous register, it remains that potential considerations on the voice – in connection with sound as distinct from sense, for example, or on all the modalities of intonation – can only be inscribed in a Lacanian perspective if they are indexed on the function of the voice as *a-phonic [a-phone]*, if I may say so. This is probably a paradox, a paradox that has to do with the fact that the objects called *a* are tuned to the subject of the signifier only if they lose all substantiality, that is, only on condition that they are centered by a void, that of castration.[20]

In a similar fashion, Dolar brings us, first, to the *aphonic* voice:

> [I]n order to conceive the voice as the object of the drive, we must divorce it from the empirical voices that can be heard. Inside the heard voices is an unheard voice, an aphonic voice, as it were.[21]

And then to its link to the Other:

> The voice is the element which ties the subject and the Other together, without belonging to either, just as it formed the tie between body and language without being part of them. We can say that the subject and the Other coincide in their common lack embodied by the voice, and that 'pure enunciation' can be taken as the red thread which connects the linguistic and ethical aspects of the voice.[22]

And here, thus and to our surprise, we arrive at the humming that sits on the other side of its topological spot, hums that are *completely and irreconcilably* the Other. Being *aphonic* sounds, these hums are the resonance of emptiness, hollowness, void. There is no human intention or imagination (though we end up helplessly imagining things from them), no signification, but 'pure enunciation' of the Other. These are hums of impersonal objects, man-made or natural things, the hubbubs that keep resounding regardless of human actions or intentionality. These hums are at odds with the personal, intimate, human humming, posing themselves completely devoid of the former. Sometimes they are resonating hums from ancient and medieval acoustic spaces like caves, ancient chambers, caverns or cathedrals, inspiring awe and immersing people in spiritual rapture, and sometimes they are nuisances, such as electrical hums at home, studios or concert halls, which need to be examined and eradicated. In other times, there are hums that can drive hundreds of people into madding distress, as mass media have reported on the Bristol Hum, the Largs Hum, the Tao Hum and the Windsor Hum, and yet the cause for such a collective experience is not always clearly known. And then there are extra-terrestrial hums: the hum of the universe, of the solar system, of Kepler 444, and so on. What is most striking to me is that how, as we were following hums directed inwards, we ended up with the hums of the other, which we assume are placed completely outside, out of ourselves. Perhaps this is emblematic of hums of the other; they are inherently directed into us; perhaps their topological spot is inside us, but we have no other way to perceive them except placing them outside, pushing them out. Why is this so? We will have to examine this in more detail later in the following chapters. But we can at least posit this conundrum by directing our focus on the association between humming and hums, which I have been making in this chapter.

Hums or humming

This association between our intimate humming and these inhuman hums appears to be arbitrary, but it is not, at least for three reasons. To start with, the 'hum' or 'humming' is used in this study as a keyword, a conceptual means that engenders a range of meanings through its signification and relations, based on which to 'expand sound studies by knowing and saying more about what we mean when we reference sound, and becoming more reflexive about how its meanings are positioned within a range of interpretations.'[23]

Secondly and more specifically, despite (or because of) the marked difference in the everyday use of these two terms – not merely what they refer to, but crucially at what register of the social, emotional, political, physiological, conceptual, and even clinical or religious structure they provoke such reference – the study uses the 'hum' and the 'humming' interchangeably. By commingling them, the study aims to highlight abrupt and sudden realisations of the common features of the two terminologies caused by the cross-reference once they begin to be examined. Interestingly, the dictionary definitions of a hum or humming offer a surprisingly differing range of meanings and connotations.[24] To hum, in its most casual definition, is 'to make a low, steady continuous sound like that of a bee', which does not seem to specify whether the agent is a person or thing. The issue of agency is mostly clarified by the first of the secondary divisions of that definition: to hum is 'to sing with closed lips' – which offers, in its direct sense, a person as the agent – while it is still vague with the second: to hum is to 'be filled with a low, steady continuous sound.' So the former definition constructs a hum through a being as it identifies who the agent is and what he or she does whereas the latter conducts it as a state, a condition, a becoming as it describes what happens with the hum. A somewhat informative definition of humming as a verb follows the latter definition: to hum is to 'be in a state of great activity,' which seems to extend the image of becoming. A subtle, yet decisive separation between the two definitions is this: the former, a definition of hum as a being, clearly identifies a person – with the body, and more crucially, the mouth – as the agent, and as such, its action – humming – is unequivocally assigned to the person; the latter, a definition of hum as a becoming, however, cannot be identified to an agent. What's more, it seems to allude that there may be more than one agent, a multitude of agents, that contribute to the becoming. Not being able to determine who or what the agent is, neither can 'to hum' be defined as a decisive action nor can it be assigned to the body or the mouth. And no agent is shown when attempts are made to define a hum as a thing: 'a low, steady continuous sound' or 'an unwanted low-frequency noise in an amplifier caused by variation of electric current, especially the alternating frequency of the mains'. I argue that this indecision of the agent is emblematic to anomalies of humming because at the most fundamental register of its being is the sound of which source, action, producer, and even listener are all obscure and ever fading away. In this regard, finally, and most importantly, even the most mundane and intimate hums that we regularly perform bear essential elements that constitute the hums of the Other, all of which are established on what can be called an ontological condition of humming – *secrecy*. And as an ontological condition, the secrecy of hums is constituted, on the one hand, by the mouth closed by the *lips sealed*, and on the other hand, by the co-presence of (or perhaps, the antinomy between) the *acousmatic*, a condition or situation in which you hear a sound without its source, and the *aphonic*, a condition or situation in which the voice loses its sound, both of which are the result of the former – the sealed lips – literally and analogically. Just as a voice in a 'signifying chain' that cannot be ascribed to the subject is severed from the chain and thrown into the Other (following Miller[25]), hums emerging from the lips sealed, in secrecy, without a shape and from nowhere, are always in danger (or in for a treat) of attracting the Unknowable because '[t]he voice [of the broken piece of a signifying chain] appears in its dimension of object when it is the Other's voice.'[26]

To follow the various shapes of hum or humming briefly examined above is to turn around our nominal assumptions about hums so as to bring their materialities and nuances to pass. The hums of the former – i.e. a hum issued from the sealed lips – assume all of the materialities of the mouth as proposed by LaBelle,[27] but the mouth is hidden and its participation in the act is passive, except, perhaps, that of the lips. However, even the lips in this case partake in preventing the other parts of the mouth from acting out. Thus, the issuing of a hum necessitates all the materialities of the mouth being negated. Furthermore, the materialities of the hums of the latter – i.e., a humming issued from the other, such as electric hums, hums of a building, mechanical hums, etc. – are completely that of the other, which is to say that their materialities are suspected because they always take the form of absence; you do not see their source, origin; they emerge not to be seen, but to be present and heard. It does not involve any of the materialities of the mouth, but one can argue that there is a 'mystical' sense of the presence of the mouth, a *spectral* mouth, if you will, which asks us to stay put and listen, and only listen. Humming in this case is called as such due to two peculiarities through which we hear it: we hear it as it resonates, acts out together with, our listening body, and as a consequence, we hear it as it connects to us, our self, *irrespective of our willingness*.

In any case, the discussion of the materialities of humming and their nuances is an extremely complicated endeavour; it requires us to open up and be sensitive to how hums couple and decouple various and differing materialities of the *real or spectral* mouth as they appear to us. On the one hand, hums originate from our body to the world in which it is situated, or they are issued from our community or society to us; on the other hand, hums do not travel outwards, but dwell on and into the subject, and through the eerie mechanism of alienating and severing the assumed link between the self and the voice, they point to the other. All hums are initiated at first as a glance at what is absent and end up pointing to the return of the other, either as a void or as a silence. What seems imperative for me is to examine whether and how these materialities connect up with our listening.

The secrecy – the hidden topology – of humming requires me to devise an unusual methodology to disclose, and uncover its nature and properties that are in the guise of absence or lack. Then, what does this methodology look like? How can one approach humming when its essential feature is to aspire to be silent, and its workings, out of sight? How can one analyse humming and its performative act of hiding? The only attribute that sets humming apart from silence – being mute – is that even with the mouth closed, it still makes a hum, the air from the lungs still excites the vocal fold, still pushes the tremors up to the mouth through the larynx and brings about just enough energy that comes to resonate in the mouth, through the nose, with the head, which then bounces back to the vocal cords and to the chest. LaBelle argues that sounds are 'micro-epistemologies,'[28] accumulating bits and pieces of the world (including the world of the subject's body) through their course of reflections, distributions and diffusions; numerous sonic rays of the sounds travel back and forth and bring with them the experiences of that world that they have occupied. They tell you a story of the world, but in a peculiar way that we need to know how to take heed of. For humming, thus, I need to find out where they hit, bounce, wind up, twirl and swirl, resonate and dissipate. Even with the mouth doing nothing much, I need to take a look at it to get to hums. So we start our investigation of humming with these conditions that establish its secrecy, beginning from the mouth and the lips to the air that they take, hold, and release. From there, the discussion will move from hums that are personal, intimate, familiar and welcome to us, to those which are mysterious, strange, eerie, and extraordinary, and ultimately – an attempt that is akin to an aim, not a goal (see Žižek for the difference between aim and goal used by Lacan)[29] – towards hums that belong to the Other. Along the way, I will try to be mindful of various shades and transformations of hums that charm, tempt, deceive, or threaten us as they resonate in and with us. I will revisit John Cage's infamous performance-writing piece, *Lecture on Nothing*, to examine the potentiality of hums that point to and connect up with a void, a silence, the other. At this point, some examples will be given to highlight what the sealed lips as an image would connote, through which I argue for the ontologies of hums. I will also take a couple of detours to stories by Calvino, Kafka, and Tarkovsky to explore the nature of hums in the guise of furtive listening, mediation, the acousmatic, and the mute.

Cognisant of the voice Lacan, Žižek and Dolar have investigated, i.e., *objet petit a*, or the object voice, I propose hums as a conjugated form of silence, which Žižek appointed as 'the object voice par excellence';[30] despite having a vocalic presence, I argue that hums are subtended by silence and its negating process. You could say that my take on humming would be as similar as that of Lagaay who, aiming toward a philosophy of voice, surveys a 'reconstruction of the place of voice in the history of psychoanalysis' by moving 'from a "positive" account of voice terms of presence and sound, to a more complex and "negative" understanding of voice in terms of absence and silence'.[31] First of all, I am not seeking to place this project within the discourse of the psychoanalysis of voice. Furthermore, although it may be true that the move from the positive to the negative, in the sense that I follow humming in its self-negating and self-negated forms and shapes, is certainly one that I am taking, it is never a move simply from presence to absence, or from sound to silence. On the contrary, as we traverse the terrain of humming, its *topological spots* – from the surface of the mouth to the dark, resonating space of the throat and the lungs, and to the air and the Other – we will soon realize that the paths are not as straightforward and smooth as such, for in the course of negation, neither does humming evaporate into silence nor its presence dissipates; instead, as Alvin Lucier's *I am sitting in a room* has sublimely demonstrated, the process is loud, resonant, chaotic and unpredictable. The spaces of absence would be full of maddening and rapturing sounds, and the presence of silences would be anything but nil.

2

The Secrecy of Humming

The lips sealed

At first, it seems easy, a passing task (it's *just* hums) to tell you what a hum is and how it makes a sound: to hum is to sing with closed lips, and you simply think of a tune and hum it. Now, though, that is too easy, too quick, and does not say much about the procedure, what's really going on with humming, so I need to try it again. First, you will think of a tune to hum, and then, you will close your mouth, draw air into the lungs through your nose – *or is it the other way around, breath in with my mouth and nose and then close the mouth?* – You hold the air for a while, think of the tune again, particularly the first few notes, while making sure you have your throat, the mouth, your nose prepared. *Come to think of it, my whole upper body is being ready for this act* – in check and under control – *do I close my eyes as well? What are they doing now?* When you are ready, start the hum – *wait, is this really what happens when I hum? Was I really planning to hum that tune? Where did it come from? Did I know I was going to hum, or did I just realize that I had begun humming? What was this tune, if you could call it as such, do I know this song, or I was just making it up as it happened?*

To me, to you

Humming brings to me one simple question for which I have two immediate answers, answers that I hope will lead us to the secrecy of hums, its structure, its workings, and its aim, once we tease out its seeming simplicity. So here is my simple question: Who am I humming to?

To me: why do I need to hum to me? What is it that makes me need to hear it? Wait, what are you really doing when you hum, are you performing this hum or listening to it? Are you doing both? *Am I in need of care, love, and a feeling of security? Am I in a shower? Am I in need of a distraction to avoid or get away from thoughts that resist erasure? What kind of thoughts do I have which I need to have erased? Why are they creeping up on me, so much so that I need a distraction? Why do I feel they are dangerous, and I have no way of keeping them under my control? Why can't I name them? Is my humming helping me remove those thoughts? Are my hums working, or in vain, useless? Am I really humming to me? What am I humming at? Is this question meaningless? Where does my humming go?*

To you: who are you? Are you someone close to me? My baby, son, daughter, grandson or granddaughter, my mother, father, my husband or wife, my love? Why am I humming to you? Do you need to hear my voice to calm yourself, (re-)assure yourself of my love, my commitment, our relationship? Or is it me who needs to hum to you? To calm myself, (re-)assure myself of your love, your commitment, our relationship? But only my hum, only my voice; not my voice proper; not my words, my reasoning, my rationale, the voice-turned-to-language, my thoughts, thoughts in chain, chain of signification? Only my air belching out from the lungs, the larynx, the vocal fold, the vocal tract? Only the guttural noises, heavily muffled, filtered, moulded and smoothed out with the lips sealed so as to bear only the traces of and from the past, assuring and reassuring, promising to commit, but not a commitment in contract, in words? Non-committal, hum and haw, fleeting, significant, but non-signifying vows, for my lips are sealed, my mouth closed, not wanting to or unable to give my words? Is this why my hums are round and round, ever more lingering, so light, floating, in the air, as the air, but never getting to you, to us? Then what am I humming at – have I asked this question before? Where does my humming go?

Hums are elusive, hard to reach and grasp. It is simple and easy to engage in them – i.e., to hum, to hear or listen to them – they require me either to simply close my mouth and let out the air from the lungs or to merely pay attention to them as they are always already here, somewhere, for me to hear and listen out. Describing and making sense of them, however, is not without effort and some negotiation in thinking. And what a negotiation it is to talk about them! Having been initially considered an insignificant task, talking about hums is becoming an increasingly insurmountable labour. There is a primary feeling of frustration you face when trying to speak of hums, something akin to that of slight suffocation. This is because they are hidden; they are behind the closed lips, and only when they are hidden, covered, and out of sight can they be made into sound and heard. For this reason, it is, in a sense, absurd to *speak of* hums that are constituted by you as being mute, *unable to speak*. What's more, their demand on the mouth – having it shut and stay closed – is unreasonable, nonsensical, and even dangerous because our mouth does need to be open, to engage, and to speak for us to live; it needs to be in action and movement, which are the mouth's *raison d'être*, through which we become its poiesis. 'The mouth,' LaBelle thus explicates,

> is an extremely active cavity whose movements lead us from the depths of the body to the surface of the skin, from the materiality of things to the pressures of linguistic grammars – from breath to matter, and to the spoken and the sounded. Subsequently, I would highlight the mouth as an essential means by which the body is always already put into relation.[1]

Instead of asking it to work harder or even merely to function properly, hums make an unspeakable request; they command the mouth to stay put and be quiet. By doing so, hums are muffling up the mouth, clearing it away, cutting the circulation of air to and through it, fettering it and effectively short-circuiting its performable apparatus. This action allows the hums to avoid the otherwise inevitable confrontation with the mouth's key pieces – the teeth and the tongue and its muscular juggling, moving to and fro, a sequence of meticulously calculated and balanced activities that they would have had to conduct. To do what? To sound, to communicate, to express, to love, to hate, to say, to say otherwise, to digress: in other words, *to mean*. Hums demand the mouth and its structure to shove over. What is this shoving over, what does it clear away? What is pushed aside? The voice? No, interestingly and unexpectedly, the voice is *still there*; it's *here*. The voice persists, even when the mouth is closed and the lips sealed. It is gaining power, momentum;

paradoxical. But this paradox, the antinomic condition of the voice outliving the mouth, despite, or perhaps because of the mouth betraying the voice, brings about a hum and sets it in motion. (As we will examine soon, this persistence of the voice, its resistance to its own disappearance, despite its decoupling from the mouth, will eventually lead us to the essence, the true nature of the mouth and that of the voice, and thus, of hums.) It's not the voice that is shackled by the sealed lips; rather, it is the voice's nominal associations with the mouth that are being questioned. Perhaps the voice does not come from the mouth as we have been led to believe, but rather it is coming from a deeper place than the mouth, which is only there at the gate and assumes ownership of its voice. This shackling, chaining the mouth up to itself, confiscating its freedom endowed from the birth of the subject, effectively puts the role of the mouth in doubt, and this doubt is in fact a conjugated form of a direct attack on language that is eventually being chained up by the sealed lips, and along with it, all its networks of 'signifying chains'.[2]

It is in this sense that, I argue, what becomes at stake with a simple act of the lips sealed to hum is to suspend and question the contracts that have been sealed between the mouth and the voice, the voice and language, and language and the mouth, both subjugating and being subjugated by each other. All the terms of the subjugation are powerfully and inherently binding, under which the self and its body are regulated and controlled so much so that LaBelle had to identify and recognize the condition of such contracts before getting to the details of oral imaginaries:

> The mouth is thus wrapped up in the voice, and the voice in the mouth, so much so that to theorize the performativity of the spoken is to confront the tongue, the teeth, the lips, and the throat; it is to feel the mouth as a fleshy, wet lining around each syllable, as well as a texturing orifice that marks the voice with specificity, not only in terms of accent or dialect, but also by the depth of expression so central to the body.[3]

Atypical, though, is the way in which the mouth is being confronted by the theorizing of humming, which, in turn, causes the materiality of humming to become more nuanced than that of any other vocalic sounds. To sound hums is to put aside the mouth and its apparatus. But that is not all; not only does humming rearrange the tongue, the lips, and the teeth, but it also reshapes its tethered apparatus, that is, the nose, the throat, the lungs, the head, and the ear. These tethered organs become aligned and realigned to make a hum and resonate with it; in this way, the whole upper body becomes a resonant chamber. And in so doing, humming reconfigures the whole relation between the subject's body and listening, and consequently, the meaning and significance of listening.

Set it free, tie me up

'The voice *stretches* me,' LaBelle confesses.[4] I cannot comprehend its significance without recognizing its religious connotation, just as Plotinus, a third-century Greek-speaking philosopher, describes in *Ennead*:

> Let us speak of the νοῦς in this way, first invoking God himself, not in spoken words, but stretching ourselves out by means of our soul in prayer toward him, since this is the way in which we are able to pray to him, alone to the alone.[5]

It is a confession because the stretching is often not 'my' doing. Rather, the voice, the sounding that is supposed to be mine, ends up pulling and bending me, wrapping me up and reaching out. It is a creature whose plasticity overwhelms the rigid structure of my body. On the one hand, this confession hints at the strangeness of the voice and its stretching of the self. We barely know its workings, and become surprised and sometimes flabbergasted by its effects on us. Hence, it is more of a question calling for an answer: *how* does the voice stretch me? This 'how' is not just directed at methods, workings, and the mechanisms of the voice; it is an inflection of *how come*, *why*? Moreover, the declaration of 'the voice stretches me' is a question of potentiality: *how far* does the voice stretch me? Asked in this way, it is not only about how far the self is being stretched by the voice; more significantly, it is in what direction does the self become stretched. Finally, it confesses that it is not the mouth, but the voice that does the stretching, which can be taken as a conundrum as it suggests that the voice may not be directly tied to the mouth. It is possible that the voice may be untied from the mouth and leave, fly away; however malleable, flexible and powerful it may be, the mouth stays with the body in the end. And thus, it is, in truth, a confession that the mouth shackles me. Notwithstanding its great performability, the mouth cannot stretch me while the voice can. And it is humming that dramatically demonstrates this. With the mouth withdrawn from its role, with its lips sealed, humming emphasizes all of these questions and painfully reveals the limits of the mouth when it comes to its assumed control of the voice. Humming, as a vocalic sound that is more closely related to silence than any other voices, is a vivid presentation of how a voice can stretch the subject, how far it can be extended, how come – or why – it allows itself to be elongated, and in what direction it does so.

Centred on the mouth – which we now know only plays the role of hiding itself – the stretching of humming is both pushing and pulling. First, humming, directed at the subject, pulls the subject inwards. When stretched inwards, a mode of humming which appears to be better known to us – '*Who am I humming to? To me*' – humming becomes a personal act of both issuing it and listening to it. It is a secretive, and sometimes mystic, act of calming oneself down and below, conjuring up memories, either consciously or subconsciously. It is a calling of the self, for the self, to hear the hum being sounded. That humming is conditioned by the closed mouth, which signifies both silence and fasting, makes humming a perfect act to bring out self-observation and control. Therefore, it is natural that various forms of humming have been incorporated into a method for spiritual and religious practices where hums are used to connect the self to his or her own body, to others, to the world, and to God. *OṃĀḥHūṃ*, for instance, the three-part mantra corresponding to the Buddha's body, speech and mind respectively, which is recited in Indo-Tibetan Tantric Buddhism for the final blessing of offerings, is considered a variation of humming where the mouth is given a minimal movement to open up, take air, and then hold it so it resounds and resonates with the body:

> From the continuum of Emptiness Āḥ (appears); from it arise very vast and wide skulls inside of which are Hūṃ. From (their) melting arise water for refreshing the feet, water for refreshing the mouth, water for welcoming, flowers, incense, light, fragrance, food and music. Appearing as offering substances, their nature is Bliss and Emptiness. As objects of enjoyment for the six senses they function to generate special uncontaminated Bliss.[6]

Such a humming practice can also be utilized to probe one's immediate surroundings – his or her physical and social contexts that can be called into question by hums – just as they can disclose the subject's relationship with his or her own body as in spiritual or religious practices. This is where hums are not pulled inward calling for the self but pushed outward to survey the world. If and when this idea is explored further, hums can lead to a study of the intimate spaces, such as the bedroom, the bathroom, the study, even phenomenologically re-constituted spaces in public space, which may emerge when one hums in the midst of a busy commuting train or bus, in the calm woods or in an open sea, among others.

Sometimes, humming is not spiritual or religious; rather, it is there to fool others who hear it, and even deceive the self who hums. It may connote an insincere or malicious intent, a plot or emotive object used in films, novels and games, such as the chilling nursery rhyme the killer hums to himself in the 1970 cult classic film *Cruising*. Further examples of this can be found in the tune hummed by the killer in *Nightwatch* or the lullaby hums by Anna,[7] the eighth killer called the Huntress, who is known for the notorious hums that make game users insane in the horror game *Dead By Daylight*.[8]

A study of humming is a complicated and complex endeavour. It is immediately a phenomenological exploration of the voice, the mouth, the body and its immediate surroundings, as well as its physical context. But at the same time, when it is stretched outward, it asks for a study of socioacoustic and sociopolitical issues. Moving from personal, private matters to a public space, hums become elevated to social and political gestures. Therefore, acknowledging that the voice stretches the self helps us formulate our questions on humming so as to ask where and how far humming stretches the self and what the possible and potential consequences of this simple vocalic act are. To answer these questions, we need to return to the mouth.

The mouth shut

Among the various parts of the mouth affected by humming, it is the lips and the tongue on which humming's demand on the mouth has the most crippling effect. Both of them are designed to be highly movable and flexible, smooth and sensitive: the lips, a soft, visible, highly sensitive part of the mouth, need to be supple enough not only to have the mouth take food, but help it produce vocalic sounds, articulate words, and compose various facial expressions and nuances; and the tongue, a muscular organ that contributes mostly to verbal communication through its movement, must be nimble enough to help swallow food and clean the teeth as well as to shape the cavity of the mouth. It cuts, divides and composes phonetic segments to construct a symbolic sequence of verbal-linguistic meanings. In addition, designated as objects charged with various sexual, linguistic, symbolic, and imaginary connotations and significations, the lips and the tongue are equally sensitized to and quickly react and revolt against any attempt to police, control, or undermine the powerful qualities ascribed to them, suggesting that they are closely tied not only to the verbal and linguistic, but to the sociopolitical systems of human beings.

The conflict between the linguistic, verbally communicative, and sociopolitical structures developed over the evolution of mankind, and the two mouth organs that these structures aim to control, is particularly telling. The so-called Pinocchio Effect[9] is a testament to the long-held belief on the ill-intending and malicious roles of the lips and the tongue, as the saying, 'his lips are moving,' may well tell us. The moving lips are deeply embedded in the psyche of human beings and communicative studies have shown, for example, how lip-reading activates the primary auditory cortex of the brain.[10] Particularly, the pervasiveness of the featural/gestural vowel information in speech that we use as cues for detecting (even illusionary) verbal sounds is so strong that even the slightest misarticulation or a-synchronisation between the voice and the movement of the lips can cause disturbing effects. As Fagel shows, when visible lip movements are presented with audible noise, the listener hears 'audible speech in spite of absence of an audio speech signal.' Following a study by Schwartz,[11] he argues that in his test, the time-variant, visible movements of the lips affected the added noise by filtering it 'time variantly.' The resulting 'perception of a time variant audio signal' produced 'the illusionary auditory perception of a spoken word.'[12]

In religious context, the lips and the tongue were the particular objects for lying, deception, and worldly desires, possible reasons why we hear in Hebrew narrative that one of the angels burned the lips of Isaiah before the great prophet heard the voice of God:

> Then said I, Woe is me! For I am undone; because I am a man of unclean lips, and I dwell in the midst of a people of unclean lips; for mine eyes have seen the King, the LORD of hosts. Then flew one of the seraphim unto me, having a live coal in his hand, which he had taken with the tongs from off the altar. And he laid it upon my mouth, and said, Lo, this hath touched thy lips; and thine iniquity is taken away, and thy sin purged. Also I heard the voice of the Lord, saying, Whom shall I send, and who will go for us? Then said I, Here am I; send me.[13]

The New Testament also considered the tongues, connoting human languages, to be useless:

> Though I speak with the tongues of men and of angels, and have not charity, I am become as sounding brass, or a tinkling cymbal. And though I have the gift of prophecy, and understand all mysteries, and all knowledge; and though I have all faith, so that I could remove mountains, and have not charity, I am nothing.[14]

Christianity had a long and contested history with the lips and the tongue. Silent prayer, for example, which had never been a common mode of prayer practice in antiquity, received a 'complicated and hesitant' reception in Greek and Syriac patristic and monastic practice in eastern Christianity. Bitton-Ashkelony notes:

> Certainly, 'silence' (σιγή), 'stillness' (ἡσυχία), and non-vocal prayer are complex categories in philosophical and religious thought and practice. Indeed, the absence of sound is one of the most obvious features of silent prayer and one that is easily discerned. Yet silence is not an absence, and it is not perceived here as an inability to conceptualize. Therefore, silent prayer, a sort of an inner communication with God, is not merely a posture before the divine, a tribute to not speaking, or a sort of mystical language of unsaying. Rather, it is a mental state in which one, through the movement of his thoughts, communicates with God and uses the concept of silence to give utterance to the essence of this interaction. The faculty that prays a silent prayer is not limited to the mind, as in pure prayer; rather the whole self is involved.[15]

But even though silent prayer was more common in Gnostic practice, the unsaying or being silent was never easily incorporated into the early Christian milieu. In Greek religion, silence or stillness was taken to be part of 'the practice of magic, curses, and petitions of criminal, erotic, or sexual nature.'[16] The deep distrust of silence – the unvoiced or the mute – which, fundamentally, was the distrust of the self, prevailed until, as Bitton-Ashkelony points out, there arrived 'the new perceptions of the self,' which would require 'new techniques by which man seeks to encounter the divine.'[17] This, in my view, explains a wider incorporation of silent meditations and related techniques in Eastern religions or spiritual practices which did not pose the self in a dualistic worldview where the self is always suspected for evil deeds, prone to be deceived and in direct opposition to the divine. The fear or anxiety that the tongue manifested evil or wrongdoing – a belief that ran so deeply into the core of Western thought – culminated in the thirteenth- and sixteenth-century inquisitions in Europe. During this time the tongue was subjected to horrible tortures like 'Branks', a cage-like helmet worn by scolds, liars, or blasphemers for public humiliation. This device was equipped with a 'very small pair of tongs that would hold a person's tongue stretched outward,' so-called a Tong Stretcher. Another method was 'Pulling the Tongue', a torture that would place the person's tongue in an iron hoop or press that was then attached to a chair with a mouth opener. In this instance the tongue usually swelled, choking the victim to death. The 'Tongue Tearer' was another barbarous form of punishment, severing the person's tongue into 'two or several ribbons.'[18] A 'Scold's bridle', a device subjected mainly to women in the sixteenth century in England and for a more extended period of time in Europe, consisted of a gag with a fearsome-looking steel mask in order to reprimand any woman who had been accused of gossiping or quarrelling.[19]

Owing to their primary condition for pushing aside and shoving over the mouth and its key parts, the lips and the tongue, hums engage in all of the above discourses and practices when we unwittingly let out gentle tunes, regardless of whether or not we are cognisant of them. But the question still stands: why do hums require the lips to be sealed, the mouth shut, and the tongue at rest? They do because hums want to make room. Make room for what? Silence.

Nothing to say and I am saying it; or, presenting everything under the light of negation

> I am here, and there is nothing to say. If among you are those who wish to get somewhere, let them leave at any moment. What we require is silence; but what silence requires is that I go on talking. Give any one thought a push: it falls down easily; but the pusher and the pushed produce that entertainment called a discussion. Shall we have one later?
>
> Or, we could simply decide not to have a discussion. What ever you like. But now there are silences and the words help break the silences. I have nothing to say and I am saying it and that is poetry as I need it. This space of time is organized. We need not fear these silences.[20]

Here, to make a case for silence, I am quoting the first two parts of John Cage's *Lecture on Nothing,* which was first delivered as a 40-minute long lecture-presentation by Cage in 1949[21] at the painter Robert Motherwell's 8th St. Artists' Club in Manhattan, New York. The same lecture was published ten years later in *Incontri Musical* in 1959, and then again a further two years later, on *Silence: Lectures and Writings* in 1961. *Lecture on Nothing* has a line, which would arguably be Cage's most frequently quoted tenets: *I have nothing to say and I am saying it.* Highly suggestive of his well-known affinity with the teachings of Zen and in a truly Cagean tone, the line seems to be lingering around the same 'nothing' as his equally infamous 'silence' piece, *4' 33"*, which was conceived about the same time as *Lecture on Nothing*, and performed in 1952 by David Tudor. 'For Cage,' Paul Sheehan notes, '"nothing" is a question of absence.'[22] He places Cage's paradoxical position of saying and nothing-to-say in the line of the 'cosmic conundrum' of philosophy: why is there something, rather than nothing, but on the other side of the question; that is, Sheehan posits that Cage's attitude toward this conundrum is not on the ontology of being, but that of no-being; nothing. It is as if, Sheehan argues, Cage were turning 'Leibniz's question around – in other words, to ask, why not nothing also, just because there is something.'[23] He, however, tinges Cage's question with a bit of melancholy:

> Cage's lecture is a lament for this lost nothing – once it is gone, it is forever irretrievable. But he is attempting, nonetheless, to recover it, knowing that it cannot be done merely through absence. As he showed, infamously, in the blank score of *Four minutes and thirty-three seconds,* 'musical' silence is not real silence, it is an enabling condition for other sonorous events: environmental murmurings and audience stirring. How can this lost 'nothing' be reapprehended, then, if not through silence? The 'Lecture on Nothing' suggests that it might be attainable through musical means.[24]

Sheehan expounds that Cage's aim is centred on his desire for 'nothing,' which is differentiated from Beckett's aim on 'nothing,' that is, to throw 'doubt on the question itself'; 'how can anyone ask about "nothing" and "something" in the same breath?'[25] I think Sheehan's take on Cage's desire for nothing or absence instead of something leads us to an interesting direction, but along the way, it has completely lost its point. First of all, the term 'desire' would not sit well with Cage's whole project if we consider the ultimate aim of his work being toward disinterest and the disengagement of one's ego, particularly that of the composer. And more crucially, Cage's affinity is not with nothing or absence, as his *Lecture on Nothing* would have led us to believe (after all, Cage wrote a piece called *Lecture on Something*). His affinity is of course with silence, but a silence that is situated both on nothing and something, or more aptly (though never precisely or assuredly) between nothing and something. What I found interesting in Sheehan's approach, though, was to do with the choice of words that he used to describe what he thought would be Cage's intentions, words like 'lament,' 'lost nothing,' 'forever irretrievable,' and particularly, 'attempting'. These words, I think, show where Sheehan's understanding of Cage's *Lecture on Nothing* is situated and issues from: in the first instance, in its initial appearance, it is on the side of the audiences, not of Cage; these words represent feelings or emotions, which Cage would have obviously not cared for much. But they are nonetheless some of the phenomena that one might have faced in Cage's lecture – if he or she had endured Cage's cunningly calculated, intentionally dull and awkward, lecture-performance, long enough with sympathies. Once placed in the audiences' minds, these words suddenly make sense: *lament, lost, forever, irretrievable,* and *attempting*, overall a feeling akin to frustration, which was well-savoured by Cage himself when he wrote the Foreword in *Silence*:

> One of the structural divisions was the repetition, some fourteen times, of a single page in which occurred the refrain 'If anyone is sleepy let him go to sleep.' Jeanne Reynal, I remember, stood up part way through, screamed, and then said, while I continued speaking, 'John, I dearly love you, but I can't bear another minute.' She then walked out. Later, during the question period, I gave one of six previously prepared answers regardless of the question asked. This was a reflection of my engagement in Zen.[26]

There is more to these words in the lecture than first appears. On a deeper level, these words in fact point to something fundamental in Cage's lecture, to the nothing that he wishes to speak of. But to discuss this, I need to introduce a few other aspects of Cage's lecture.

In a more contextualized examination of Cage's lecture, Brown takes Cage's tenet to be a more direct attack on the passion toward something to express, which was the currency of artistic discourse in the mid-twentieth century in North America.[27] This predilection was particularly prevalent in New York under the sway of Abstract Expressionism led by painters such as Pollock, Rothko, De Kooning, and Barnett Newman who, Brown argues, '"stood" for something.' Brown saw that Cage's lecture was specific to the situation and the audience to whom he was to addressing or performing. Having already thrown away the belief that music could express anything, let alone feeling or emotion, Cage, Brown posits, would have been in an uneasy situation – certainly an opportunity Cage would not have wanted to miss – when he was asked to present something – *anything!* – at the 8th Street Artist's Club, also called 'The Club' – a hotbed for New York Abstract Expressionism amongst others. The first line of *Lecture on Nothing*: *I am here, and there is nothing to say. If among you are those who wish to get somewhere, let them leave at any moment*, Brown argues, is a direct attack on Abstract Expressionism and those who believed what it stood for.

Furthermore, Brown recognizes the systematical nature of Cage's lecture, pointing to the structure and form of how it would have been delivered, which Sheehan also notes: 'a formally striking layout designed to mimic the rhythmic structure of Cage's musical composition'.[28] In fact, we can see this structure from the printed version of the lecture, the description of which reads: 'There are four measures in each line and twelve lines in each unit of the rhythmic structure. There are forty-eight such units, each having forty-eight measures . . .' Thus, the first two parts quoted above should now look something like this:

I am here , and there is nothing to say .
If among you are
those who wish to get somewhere , let them leave at
any moment . What we re-quire is
silence ; but what silence requires
is that I go on talking .
Give any one thought
a push ; it falls down easily
; but the pusher and the pushed pro-duce that enter-
tainment called a dis-cussion .
Shall we have one later ?

♍

Or , we could simply de-cide not to have a dis-
cussion . What ever you like . But
now there are silences and the
words make help make the
silences .
I have nothing to say
and I am saying it and that is
poetry as I need it .
This space of time is organized
. We need not fear these silences, —

What is it that Cage is *attempting* to do here? Brown rightly observes that this proportionally structured text will encourage the reader to 'recite the text according to a general tempo, giving pause in the spaces between words accordingly.' The *pause* is a silence, and this is the whole point: there is nothing 'musical' about this other than the somewhat awkwardly arranged and stuffed-up texts under the seemingly rigid 4/4 rhythm. It would be wrong to try *performing* the lecture as a musical piece, to fashion it as sort of a rhymed, lyrical song; instead, the focus is to aim at creating pauses, lots of them. Indeed, that is what Cage says right from the beginning – *I am here, and there is nothing to say*; in other words, I argue, there is no music to play, but I am playing it, why, because we need silence; or, there is no need for me to be here, but I am here, why, because we need silence. Brown seems to take the argument further to connect Cage's pauses, silences, and particularly, his specific usage of 'need', 'ought', Necessity, to American Transcendentalism and figures like Emmerson and Whitman. Indeed, there is no arguing on the close link between Cage's writings and those of the American Transcendentalists, which had heavily influenced Cage's philosophical thought. Christopher Shultis notes the similarity between the emblematic line from Cage's *Lecture on Nothing* and a line from Thoreau's Journal, 'The art of life, of a poet's life, is, not having anything to do, to do something':

> Having nothing to say and saying it, and doing something when having nothing to do – are these statements saying the same thing? Does it matter that the second statement was written in a journal by Henry David Thoreau in 1852, almost a century before Cage gave his 'Lecture on Nothing'? I think it does, even though if one took a strictly historical context into account a different Thoreau might emerge. Nor is it in the spirit of Thoreau, whose overriding interest was in the 'present moment,' to suggest that historical context is solely of the essence when discussing influence in the making of art. In 1989, Cage began a text entitled 'An Autobiographical Statement' as follows: 'I once asked Arragon, the historian, how history was written. He said, "You have to invent it."' In the history that Cage invented, Thoreau played no small part.[29]

As Shultis further discusses, the similarities between Cage and Thoreau with regard, among others, to their 'non dual approach to sound and silence and their resultant appreciation and acceptance of non-intention' are striking, considering that Cage had apparently become acquainted with Thoreau's writings six years after he published *Silence*:[30]

Cage composed music like what Thoreau himself heard by just listening. This is why Thoreau did not have to go to Boston to hear the symphony. For him, as with Cage, all sounds, all noises were music: 'Nature makes no noise. The howling storm, the rustling leaf, the pattering rain are no disturbance, there is an essential and unexplored harmony in them'.[31]

Disquieting silence

Having said that, I now want to take a slightly, yet, decisively different, and disturbing – and indeed, *disquieting* – view on Cage's *Lecture on Nothing*, in order to examine modes of humming. That is, I think examining Cage's lecture and his mode of presenting 'nothing' will shed light on how hums are presented to us and how we present them. Cage's lecture has long and often been less well understood, even though scholars and others have quoted it heavily. The most-quoted line, 'I have nothing to say and I am saying it and that is poetry as I need it,' has often been taken out of context, a context which would tell us the reason why Cage is saying it despite the fact that he has nothing to say. So the line needs to be contextualized by the line above it: 'What we require is silence; but what silence requires is that I go on talking'; and another after the line, 'We need not fear these silences.' Why do these lines need to circumscribe it? Why do we need consider them together? It is because they can help us get past a spate of discourses and assumptions that surround and obscure some of the less examined aspects of the lecture, which I think are not only crucial for us to understand *Lecture on Nothing*, and for that matter, most (if not all) of Cage's music, but more importantly for us, directly connected to the secrecy of humming.

First of all, let us start from Cage's original intention, presented as simply and clearly as possible in his performance text, which begins with this: *I am here, and there is nothing to say*. From the start, Cage is playing with a paradox of beings; in other words, there is a presence of 'I,' which is in conflict with *a presence of* 'nothing', and that conflict is enacted by 'have' and 'to say' as in: 'I have nothing to say'. Specifically, though, this is not a direct conflict, standing in opposite directions, meaning that 'I' does not clash with the 'nothing' that 'I' is trying to say; rather, the two are in different modes of being, hence the difficulty of saying it. I cannot dwell too much on these two different modes of being as I have more urgent tasks at hand, but it should suffice to say that the presence of 'I' engages in that of the natural mode, that is, it appears as a positive being that we do and can see and interact with by a process of affirmation, and along with it, all the actualized and possible operations; the presence of 'nothing' engages in that of the negative mode, that is, it appears as a positive being that we can interact with only by a process of negation, and along with it, all the potential and symbolic operations. Important to note here is that it is Cage who makes it appear and brings it to a symbolic discourse through his process of negation. Moreover, what Cage is intending to do is 'to say nothing,' which is worlds apart from 'not to say anything.' In this sense, like many of Cage's musical projects, *Lecture on Nothing* is a positive, active attitude toward nothing or absence; it is an attitude of Necessity, of the 'ought'; everything – and even nothing – ought to be present. Therefore, it is quite wrong to take Cage's project – this lecture and *4' 33"* for example – as a project towards negation, nothing, absence. A negative process is implemented, but the negation is not the aim. It is neither nothing nor absence that is questioned as Sheehan argues. Rather, and to our surprise, Cage does present things – everything, really – positively, and the mode in which he presents them is by negation. In other words, with Cage, everything is presented under the light of negation, but then with Cage, that process of negation is *silencing*. Kahn discusses Cage's visit to an anechoic chamber at Harvard University in 1951, one of Cage's key moments of 'silencing' things in order to present us with everything, including 'the impossible inaudible'.[32] Kahn points out how, since then, Cage had more frequently begun incorporating electrical means to amplify small, otherwise inaudible sounds. 'The anechoic chamber was,' Kahn explains, 'the technological emblem for Cage's class of silencing techniques'[33] along with other 'techniques,' like silencing traditional instruments in a social, cultural context in *4' 33"*, the 'canned silence' in *Silent Prayer*, which was proposed to be programmed as a Muzak on a radio programme, or an even more fanatical imagining of banning and 'destroying the recordings and the means for playback' altogether. More symptomatically, though, in the process of amplifying small, almost inaudible sounds and making them audible using technology, Cage started elevating this technique of amplification into a method of its own and eventually turning it into 'a discursive means for musical listening.' Kahn sees this trajectory of Cage's commitment to the amplification process as that of 'the impossibility of silence,' under which 'the world was suddenly overrun with small sounds,' and thus, even without the absolute need for such a method, Cage would still use it 'to perform rhetorically, far beyond its actual technological capabilities, to increase the number of possible sounds and to deny inaudibility.'[34] This denial of inaudibility is, of course, connected to Cage's dictum, 'there is no such thing as silence'; however, with the technology of amplification, the 'is' in 'there is' now sounds more like *ought to be*. This force of Necessity is what Kahn takes issue with regarding Cage's silencing project. But there is more. According to Kahn, Cage ends up silencing 'the social, political, poetic, and ecological aspects' of sounds because of the composer's indifference in reconceptualizing 'the sociality of sounds' even after the shift of his interest from music to social issues. This, Kahn posits, results in Cage's musical thoughts and ideas being incorporated into 'a tradition of mythic spaces' as the composer circulates 'the sociality of sounds through an impossible and implausible acoustics.' It is in this line of thought that Kahn objects

> to accept how Cage reduces sounds to conform to his idea of selfhood. When he hears individual affect or social situation as a simplification, I hear their complexity. When he hears music everywhere, other phenomena go unheard. When he celebrates noise, he also promulgates noise abatement. When he speaks of silence, he also speaks of silencing.[35]

I agree with Kahn on how incessantly and progressively Cagean sounds disappear, on the one hand, at the threshold of 'amplification' in the literal sense of him using technological means, and, on the other hand, with the way in which his approaches of silencing end up either connoting or promoting politically or socially irresponsible positions. Disappointingly, however, Kahn either quickly discounts or doesn't concern himself much with a psychoanalytical approach to Cage's silences and silencing. I also agree with Shultis that Cage is attempting to redefine 'nothing' as the void. But again, Shultis does not go on to examine that void: a void of what and whose void it is. If this void is to be that of Lacan, and when it is Cage's (or whoever's, for that matter) *voice* that (attempts to) announces that void, the text, 'I am here and there is nothing to say,' takes a startling turn. It is in this turn with which I want to take another look to the mode of presentation of 'I,' 'nothing,' 'have', and 'to say', all wrapped up under the positivity out of Necessity. In this turn, I propose that Necessity is not coming from Cage, the self, but from something else; that is, I argue, it is not Cage's personal desire or intention, but the drive or the void that announces itself.

The first thing to note: in *Lecture on Nothing*, that void is silence, or rather, *Cage's silences point to the void*, the same void bespoken by Lacan, the other. Additionally, the deafening inaudible that are 'amplified' by Cage's 'silencing' are still silences, now fully formed and incorporated into those which are presented – 'now there *are* silences.' Still as nothings, they are nothings that have appeared and can be appreciated and experienced, however frustrating and awkward, in the way they may appear to us. Furthermore, we can identify some peculiar conditions in which *his silences* appear and point to the void: the first peculiar condition – if Cage's aim is to have silences (although he also uses silence as mass noun, it should actually be a singular noun, meaning always a silence or silences, for Cage's silences are always contextual), why would he have to be present, 'to be here'? That is because for *these* silences, which point to the void, to emerge, we need an agent, an agent of silences, particularly, one with a mouth which is supposed to say 'things'. In this regard, *his silences in this context point to the object voice*. In *these silences*, what kind of 'things' one needs to or can say does not matter as much as that one needs to be here and keep saying it so the silences can emerge: thus, Cage goes 'on talking'. In this sense, *his silences need sounds*. In the third peculiar condition where the sounds need to be vocalic – not only do they need an agent and a body to be present – they also need for the agent to keep talking. In this case, what is spoken must be nothing in particular, so Cage's mouth can utter anything. He further clarifies that, while an utterance can assume the form of a discourse, a conversion or mundane activities of the mouth, it doesn't have to be. Cage says, 'we could simply decide not to have a conversation. What ever you like.' Thus, any vocalic sounds, really, any nonsensical babbling would do. Thus, *his sounds for silences do not need to mean anything*. Finally, though, there is one strict rule applied to these vocalic utterings that keep on talking – the presentation of those sounds. They need to be cut, sectioned and shoved in a shoebox of four measures in each line, with twelve lines in each unit, which can be clearly visible in the printed text. This rule was implemented meticulously by Cage and agonisingly experienced by the audiences during the lecture-performance. The pauses, stops and cuts impede, delay, and problematize the habitual flow of communication – both speaking/talking and listening/hearing, resulting in an estrangement of the words, those voices, from the mouth, from Cage, and from its audiences. Once this rule is implemented, once Cage utters words and keeps on going, 'now there are silences.'

Humming and four modalities of Cage's silences

Now that we have identified the components of Cage's silences on *Lecture on Nothing* and examined the peculiar conditions based on which his silences appear and point to the void, we are ready to delineate four modalities of Cage's silences from which I will make direct links to humming. The first two are more pertinent to our aims, thus, needing more space to examine, while the rest are lighter in their approach to Cage's project, so my explanations of them will be brief.

First, the way in which Cage gives space for silences so as for them to appear and be present follows the same process in which hums appear and are present for us to hear. Just as Cage's silences exploit the antinomic interplay between the presence of 'I' and that of 'nothing' through the operation of *having* and *saying,* hums puts the mouth (and its tethered apparatus) and the voice in the same antagonizing interplay, through which hums stretch the self in and out. The operation of hums in this conflict is driven by a strange and eerie lack of commitment from the mouth, the voice and the self. This lack of commitment, just like the lack we experience from Cage's lecture, invites hums, which were enunciated by the self initially, but now stretches the self *in and out,* overpowering it as if the hums had been there, here, all along, just waiting to have space. And when (not if) this happens – *when* the self recognizes what is happening – the *disquieting* creeps in and starts to settle.

In *Lecture on Nothing*, the modality of silence – or form in which it appears – is unique as it combines two modes. Cage's performance is not only aural, but also visual in the sense that it has been printed to show how Cage's silence appears and inhabits the space given by him. What needs to be clarified – and in this sense, Cage himself created a potential pitfall, out of his tendency toward experimentations on graphic scores – is to remind ourselves that what we see as spaces are not silences; silences become only present when *one is present and keeps on talking*, not by showing where and how those silences can be created by formulation. Thus, the visual with which Cage *ought* to present silences is really a kind of mental visualisation. Interestingly, such a practice of combining the two modes – auditory and visual – to give silence space can be traced back to ancient and medieval religious and performance practices. Aiming to answer the role of silence in Gregorian chant before the invention of the measured notation of silence in the thirteenth century, Hornby examines indirect clues like comments and implicit evidence left by practitioners within chants. These show the use of pauses either for practical purposes: *media distinction* (taking a breath between singing); for meditative purposes (in antiphonal settings, singers often take a pause and listen to what the other group sings); and for communal unity as they breathe in unison. This leads Hornby to conclude that,

> [i]n psalm singing, silence is a well-attested part of medieval performance practice, but its meaning is not primarily musical. Instead, the focus is on meditation, unity, ceremony, the visitation of the Holy Spirit, and the release of mood-enhancing endorphins through the controlled breathing of the whole monastic community.[36]

And thus, she argues that the silences of medieval monastic psalm singing demonstrate their 'transcendent possibilities, or at least the possibility of a change of mental stage.'[37] Hornby's work on psalm singing was taken as a case study by Williamson (2013) who argues that '[m]usic can be presented visually, so that it signifies silently, in the inner ear'. In doing so, it 'can also be present, but silent, in moments of pause in the midst of an actual physical and sonic performance.'[38] This is because while not sounding physically, the music has not stopped.

The singers are aware, within their own interiority, of its continuation, and though they do not hear it in their physical ears they hear it still inwardly. In this moment of silence, the music does not disappear, but functions temporarily – and temporally – on a different level.[39]

More rigorous and inclusive practices on the use of silences were employed in mentally performed music in Tibetan *Chöd* rituals, either as silent meditational performance or mentally produced music that is believed to surpass what is physically performed. Cupchik points out that the melodies and rhythms in *Chöd* performance utilize various compositional techniques, such as 'tone painting, melodic phrasing, sequences, and rhythmic ostinato.' These aim to 'enhance the meditation process by eliciting specific emotions,' helping the ritual practitioners, among other things, with cutting (*Chöd* means 'cutting through') or dissolving notions of the self.[40] He goes on to argue:

Within the context of a *Chöd* ritual, the practitioner cuts her attachment to the notion of 'self,' enacting a concept of intersubjectivity and compassion that is central to the essence of Buddhism by emphasizing the mutually interdependent nature of phenomena, and dissolving the egotistical framework that separates 'self' and 'other.' The purpose of the *Chöd* ritual is to destroy the demon of 'self-grasping ignorance' by 'cutting it off' (*gCod*) at its root. To do so, the *Chöd* practitioner is instructed to perform the ritual in frightening sites where spirits are said to live (cemeteries, haunted places, temple ruins, ramshackle dwellings, the meeting of two paths in a forest, etc.).[41]

In *Chöd* rituals, though, silent meditations are neither mental offerings taken during pauses between performances nor additional or complementary practices to the main ritual activities. Rather, silences can even surpass what is being seen visually or heard aurally:

Typically, during the oral transmission of the *Chöd* ritual, the lama will insist that music is not the central point of the practice. Rather, one should focus the mind on increasing one's level of altruism (*bodhicitta*) and attaining a deeper level of realizing the

> nature of interdependence (emptiness). According to a preeminent mid- to late twentieth-century exponent of the ritual practice, Kyabje Rinpoche, former head of the Granden *Chöd* tradition, one does not even need the music to practice *Chöd*. Ven. Pench Rabgey-la echoes this sentiment when he tells me legends of *Chöd* adepts (including his own teacher) who found it just as effective to sit in a cemetery (a 'sky-burial' site in Tibet) at night to perform silent meditation, thinking through the process of a *sādhana*, mentally performing all the visualization in each step, and in this way reaching attainments.[42]

This is in line with Ellingson who observes the visualized music or mentally produced but not physically present music in Tibetan rituals. Remarking on the quality of these rituals, Ellingson discusses what overpowers and encompasses the music that is being physically presented:

> From a performer's perspective, the whole of the music offered in a given performance is always more than the sum of its audible parts. Furthermore, not only is the 'music' substantially different from the sounds heard; it is also different in different ways for each individual performer![43]

It is not my intention, with the above references, to connote ritualistic aspects of Cage's lecture (although one can certainly draw a link if he or she follows the effect/affect of Cage's lecture *cutting through* the psyche of the audiences, those who (are supposed to be present and) listen). Rather, my aim is to show and highlight where Cage's silences are located and how they emerge. Cage brings out his silences by positing, and forcing the audiences to recognize, the ontology of our being, the *always already* established antinomy deeply ingrained in our being that assumes rigidity and stability not through its own innate structure, but through our hopeless belief in prolonging it. The irony, though, which Cage must have recognized at some point of his whole project towards silences, is that Cage's silences – and for that matter, all silences he intended and others he did not intend – do not need one's intention to bring themselves out; they have already inundated us, both individually and collectively, and our world. The (grave) mistake, I argue, Cage permitted and allowed to happen in *Lecture on Nothing*, and possibly in his whole project, was to have made us aware of that. They are in reality not hidden and waiting for us to find them. They are already here. In other words, the mistake, thus, is that Cage thought, and we believed, silences are behind everything which sounds, but no, it is the other way around: sounds are behind all that which is silent, absent. As the fundamental condition for our being and existing as the self in the world, we spare no effort in silencing those silences with our busy, small, minute, insignificant hubbubs, which have been amplified inexcusably and have become a particular symptom of our age since the nineteenth-century industrialization and the overpowering capitalism that immediately followed it. Perhaps the hubbubs of our age are the 'realities' that we employ to suppress the Real, 'an unrepresentable X, a traumatic void that can only be glimpsed in the fractures and inconsistencies in the field of apparent reality,' if we posit that these hubbubs are one of the symptoms of the incurable maladies of capitalism that Fisher points out.[44] Going back to the silences and sounds and their reversed position, Žižek thus remarks:

> as Lacan points out, voice and silence relate as figure and ground: silence is not (as one would be prone to think) the ground against which the figure of a voice emerges; quite the contrary, the reverberating sound itself provides the ground that renders visible the figure of silence. We have thus arrived at the formula of the relationship between voice and image: voice does not simply persist at a different level with regard to what we see, it rather points toward a gap in the field of the visible, toward the dimension of what eludes our gaze. In other words, their relationship is mediated by an impossibility: *ultimately, we hear things because we cannot see everything.*[45]

With this, Žižek recognizes, and makes us 'confess,' the lack or the gap (or the Lack or the Gap, to recognize their otherness coming from the big Other) and the way that we confess this through the objects; for Freud, it is through the breast, the faeces, and the phallus; for Lacan, it is through the gaze and the voice. And for Cage, it is through his presence 'I,' which he has been struggling with all his life and career; through the presence of 'nothing,' which now reveals its true nature not as *Nothing*, but as a conjugated form of 'I,' a lost one at that (or one that he wishes to lose?); through 'have,' which is the other side of the 'lost', thus forming the antinomic structure; and finally, through 'to say,' which is an attempt to gaze at the lack or the gap, which is silent. Now if we substitute these four components of Cage's lecture with something else, we have hums: with hum, we confess the same lack or the gap as that of Cage through the presence of 'I,' which is the self who hums; through the presence of 'the mouth,' which is a subjugated form of 'I' and is now receded into a dormant, sealed, lost, thin line; through 'the air,' which connotes both 'having' and 'losing'; and, through 'to hum,' which is an attempt to gaze at the lack or the gap, which is also silent. But here, I argue one crucial and significant difference between Cage's performance-lecture and hums: they both activate the gaze at silence, the same Silence; yet Cage's performance-lecture ends up not even attempting to listen to it (which is an extremely crucial, and truly troublesome, point with regard to Cage's project, but then, taken another way, one can ask: can we listen to the Silence when we face it?) and devotes itself wholly to the gaze; whereas the latter, hums, invests its whole action to listening. And in this sense, hums are the closest form of silence, not merely in the way they appear and present themselves to us, but also through the way in which they face the lack or the gap.

Second, we need to revisit the structure of this line, 'I have nothing to say and I am saying it,' in order to question the nature of 'ought' or Necessity that many have pointed out as Cage's key philosophical allegiance to American Transcendentalism or Zen. Indeed, it reads as if Cage were speaking with his unassuming, gentle voice (Cage had such a unique and disarming voice that you couldn't forget it once you had heard him talk) in full willingness and care. I beg to differ. For I am suspicious, in the line shown above, of the use of 'and.' For some, this seems natural considering Cage's affinity to the Zen practice, such as flow, openness, letting-be, 'what ever you like,' and so on. I argue that this 'and' is a disguise, as it is rather a 'but.' Now if rewritten and read again, 'I have nothing to say *but* I am saying it,' the line certainly takes a sense of urgency, a tension, a push-and-pull. Cage certainly suggests this:

> Give any one thought a push: it falls down easily; but the pusher and the pushed produce that entertainment called a discussion. Shall we have one later?[46]

What's more, he says: 'What we require is silence; but what silence requires is that I go on talking.' But he does not disclose what the reason for that requirement is. Why do we need silence? Why are you here? Why are you saying things? Why are we here, gazing at you and listening to your voice? If we continue reading Cage's lecture, the Necessity, the need, the ought, gradually recede from view, which is at equal speed taken up by anxiety, frustration, boredom, *and fear*; the fear of silences. Now he said this earlier in passing, but at this point, it reads like a warning: '*We need not fear these silences.*' Is Cage's lecture, having started as an uneventful, quotidian monologue, going to turn out to be a horror drama?

If you think my take on this line is a warning, and thus, turns Cage's lecture into some kind of a horror classic – that it is perhaps stretching my argument too far and making this particular attempt as a joke – let me assure you that it is not. But I might restrain myself and say that Cage's *Lecture on Nothing* does not evoke the sense of horror or terror – *not yet*, that is. What I want to point out is that the lecture has all the hallmarks of a typical – read, archetypal – event that insinuates that which is to become strange, weird or eerie, which can then turn into a horror or a terror: first, the way the presence of 'I' is announced – 'I am here'; second, the lack of the reason for the presence of 'I' – 'I have nothing to say'; third, the persistence of an uncalled-for action or behaviour – 'I go on talking'; and finally, *the presence of something that compels* 'I' to do that uncalled-for action or behaviour. Strangely, while we can identify the first three features, and indeed, Cage made sure that they are clearly visible, not so apparent is the final feature – what compels him to do it. Instead, we have 'require,' 'could,' 'shall,' and 'and'; at best, he hands out the 'need' – Necessity. I don't buy it. What I believe compels him to be present and keep saying even if he has nothing to say, is *a drive*, and the silence we, eventually, end up facing in both Cage's lecture and hums is *the silence of the drives*.

It is important to note that silences in Cage's lecture are in a variety of different shades. According to Dolar, silence can be grouped into three categories following Lacan's three registers.[47] In the symbolic register, something absent can 'refer to something just as well as something present.' This co-existence of an absence and a presence is characteristic of a symbolic silence, such as 'the absence of a phoneme, in order to get the meaning' in communication. Similarly, 'the absence of a sign as a highly meaningful sign' as well as the 'place of a sign, designating its absence, is itself part of the sign, on the same level with it.' In this sense, symbolic silences are deeply rooted in the practice of rhetoric, and thus, can be immediately experienced in those numerous pauses that Cage designed within his lecture. In the imaginary register, on the other hand, silence 'can indicate the highest wisdom, and its extension can be a "mystic silence," a silence of the universe which can overwhelm and enthrall the observer, a vision of supreme harmony, the oceanic feeling that Freud talks about in *Civilization and Its Discontents*, cosmic peace.'[48] In the lecture and many other talks and writings, Cage seems to engage with silence in the symbolic register, aiming to connote it in the imaginary register; his methods often play with the tension of presence and absence, through signifiers, symbols, and also phonemes, but their aims appear to be most closely linked to an ideal of Zen. However, what drives Cage to be present and say 'nothing' though he has nothing to say, and what Cage and we end up facing, is silence in the register of the real. Here, no longer does silence concern itself with the wisdom, harmony or some kind of grand scheme of the universe as it does in the imaginary register; instead, it severs its contribution to sense of anything. This denatured silence, Dolar argues, is the 'most disturbing feature' of silence:

> There is nothing natural in the silence of the drives – this is not the muteness of some natural life, it does not pertain to some organic or animal base; on the contrary, the drives present a nature denatured, they are not a regression to some originary unsurpassed animal past which would come to haunt us, but the consequence of the assumption of the symbolic order. They get hold of the organic functions and corrupt them, so to speak – they abolish the natural functions of the organs and turn them into extensions of a phantom organ.[49]

The 'assumption of the symbolic order,' which the lecture has employed and integrated into the mode of its delivery by Cage, results in these drives of the most denatured origin. Perhaps this is why Cage, unwittingly, gives us a warning that 'we need not fear these silences.' But how hopeless it is, if we are indeed in front of these silences! As I argued above, Cage's lecture and most of his projects towards silences do not really face the silence of the drives, even though they are issuing from it. Cage himself does not seem to show any responsibility for the chaos and havoc that his projects of silencing have let out; and how about the audiences, can they then feel responsible for the silence of the drives? Not really, and I think this is the fundamental problem that underlies the disquiet surrounding his projects. They establish a formula that encourages us to bring out silences, and then, takes a peculiar Cagean guise of disinterest in them. Hums, on the contrary, following the same procedure of 'gazing at' and 'voicing' those silences, end up having us face and confront it. The method of such an audacious act – although it sounds circular and simplistic – is to keep on humming. How is this any different from Cage's urge to keep on talking? The crucial difference is that to hum is, by its very ontology, to hear and listen to what is being hummed. Perhaps, this may be the only way to face the silence in the Real as Žižek notes: '"Why do we listen to music?": *in order to avoid the horror of the encounter of the voice qua object*,'[50] because 'Hearing voices in that most improbable but tenacious and insidious way is hearing a silence.'[51] In the following chapter, where I will examine the hums of the other, I will discuss in more detail the malicious and destructive affect of humming, which is taken to be the repercussion of facing the silence of the drives.

Third, I would like to observe a similar approach made by Cage in *Lecture on Nothing* and humming towards silence, particularly, aiming to note the simplicity with which both invite silence and the void. With Cage, it is as simple as him being present and giving pauses between whatever he says. It is in the same simplicity that hums come to us; they come unannounced and are already with us; they have always already come. Just as Cage starts with his lecture, 'I am here,' the only action needing to be taken is to hum. This swift approach to silence reminds me of *lightness* and values, qualities that were dear to Italo Calvino, a great Italian storyteller. Calvino thought that these qualities would be relevant to twenty-first century literature, but they may also be useful for other practices, including what I am about to discuss. To talk about the values of lightness in opposition with weight, Calvino tells a story from Ovid's *Metamorphoses* about Perseus and Medusa to illustrate the balance he needs to strike 'between the facts of life that should have been my raw materials and the quick light touch I wanted for my writing':

> To cut off Medusa's head without being turned to stone, Perseus supports himself on the very lightest of things, the winds and the clouds, and fixes his gaze upon what can be revealed only by indirect vision, an image caught in a mirror.[52]

Then, Calvino goes on to say:

> As for the severed head [of Medusa], Perseus does not abandon it but carries it concealed in a bag . . . It is a weapon he uses only in cases of dire necessity, and only against those who deserve the punishment of being turned into statues. Here, certainly, the myth is telling us something, something implicit in the images that can't be explained in any other way. Perseus succeeds in mastering that horrendous face by keeping it hidden, just as in the first lace he vanquished it by viewing it in a mirror. Perseus's strength always lies in a refusal to look directly, but not in a refusal of the reality in which he is fated to live; he carries the reality with him and accepts it as his particular burden.[53]

A similar, heavy and thorny question that Cage may have grappled with would be this: how can 'I' be with 'nothing,' let alone have 'nothing' and 'say' it? Cage then does something that is akin to what Calvino highlighted in the way Perseus dealt with the head of Medusa: he presented this paradox lightly and swiftly as if there were no paradox at all; and once presented as such, it now becomes the burden of those who have witnessed the emergence of his conundrum. In this kind of lightness and simplicity, I argue, humming invites the void. The unspeakable or mute conjured up by the shut mouth and sealed lips and the gentle hums that they let out are nonetheless analogous to the terrifying head of Medusa and Perseus, who glances sideways at the horror via a mirror, handling the decapitated head with care and gentleness, yet still hidden.

Finally, just as in Cage's lecture where the semantic chain of words and sentences are interrupted and mangled by gaps and pauses so that their significance and meaning become not just nuance but also often completely dissipated, hums also lose their link to any linguistic framework the voice may have. In fact, hums slip through the net of the conceptualisation and the discourse of the voice. It does so by hiding (sometimes unsuccessfully) the almighty text – the lyrics – that in turn smooths out what Gumbrecht termed 'the meaning effects'. The same process of 'smoothing-out,' though more forcefully due to its rigid structure and rules, occurs in Cage's lecture. What both hums (if they can intend) and Cage's lecture aim at, if not immediately or deliberately, is, of course, to have us experience that which has come to the fore of our perception because of the mechanics of silencing, which may bring out what Gumbrecht called 'the presence effects.'[54] But, such an act of a 'smoothing-out' of meanings sometimes brings back the signifying system more intensely if and when the text returns with a vengeance, much like a pendulum whose force of returning becomes greater after it had been swung away. In Cage's lecture, we cannot help but seek for meanings or significations despite the impeding stops and nonsensical composites of words; similarly, we also cannot help but search for significance as we hum or hear them. This is a reason why, perhaps, Cage pointed out that 'something and nothing are not opposed to each other but need each other to keep on going,' in his equally challenging performance-lecture, *Lecture on Something.*[55] Gumbrecht posits that our aesthetic experiences swing between these two effects, recognizing the tension generated by the complicated processes of the meaning effects and presence effects.

But there is one element in humming that sets itself apart from Cage's lecture with regard to the workings of the meaning effects and presence of effects. The silences that we face in Cage's lecture, which eventually point to the void, always swing between the presence and meaning effects, and by so doing, they are here at present and can be experienced, mainly due to the fact that there is a continual tension exerted by the presence of 'I' – Cage – and his continuing speech process. In hums, however, the presence of 'I' is from the beginning vague and prone to disappearing, even with the continual humming. The hums that keep on humming do not always evoke those who hum or those who listen to them here and now; on the contrary, there is a tendency (or a danger?) for everything – from the person who hums, anyone who potentially hears the hums, and even the hums themselves – falls into a void, whether that void consists of individual or collective memories, a past or pasts, a future or futures, and whether humming is out of care, love, melancholy, longing, mischief or deceit. All seem to recede into the past or the future. This is because, in hums, the pulling of the other, the force that the aural materiality of sounds that the shut mouth and sealed lips produce, is often too strong.

And that pulling is responded to forcefully by the meaning effects, which aim to turn every instance of that pull from the other into signs. Thus the end result becomes, somehow, tinged with a sense of those which have passed or those which are to come – wishful, hopeful, whatever – but never those which are actually present. To explain it, we need to continue following Gumbrecht's investigation on the conundrum of these two effects. An interesting example he gives to highlight the implications of the presence-meaning pendulum is the transition of the Eucharist from the medieval to Protestant theology.[56] The strong connotation of magic and mystical qualities that the Eucharist of the pre-modern and Catholic Eucharist made via the 'substantial presence' of Christ's 'body' and His 'blood' meant that the Protestant theology had to turn these presence effects into some kind of 'evocations' and 'meanings.' Through this transformation of the presence effects, Protestant theology developed its own 'commemoration,' first conceptualized by John Calvin. This decisive transformation of presence effects to meaning effects in the Eucharist, Gumbrecht argues, was also a departure from the 'temporal distance' that mass participants might have experienced via the enactment of sharing the 'bread' and the 'blood'. Indeed, this is akin to the 'historical distance' that church-goers now experience with a commemoration of the Last Supper, which now comes to them only as a sign or memory. Here Gumbrecht points out that in the transition from the experience of materiality in temporal time to signs of materiality in historical time, 'signs at least potentially leave the substances that they evoke at a temporal and spatial distance.'[57] Following him, we might say that hums, once turned into signs – which is to say, those which can be turned into signs, those of whose meanings we are able to decipher, those that are gently swinging to and fro between presence effects and meaning effects, those which are present in the layer of the symbolic register – can be rather easily incorporated into our being, our life, our world. We know those hums; they are familiar to us and they are intimate to us, even though they are sometimes slippery and difficult to grasp. We are comfortable with them, in them, as we are the actor of those hums. Even though we may lose things because of them, we also gain things with them. What are we losing contact with by humming? Quite a lot: language, a majority of vocalic expressions, taste, the inside-outside network of the mouth through which we breathe, speak, sing, shout and scream, cry, eat, spit, vomit, cough, burp, choke, suck, lick, stutter, and whistle – almost all the oral imaginaries that LaBelle examined.[58] What are we gaining contact with by humming? A few, but significant few things will resonate with hums: our body, which you can feel vibrating, our hearing/listening, our world (which includes one's immediate surroundings and spaces and), a passage toward our memory, and a channel that opens up and connects with others. In this sense, humming is the very act of connecting to our world and to our self, 'the vocal link back to the corporeal.'[59]

And this is it. This is all I can do. This is the end of the push; now I am being pulled back again. The string that holds me, which swings me back and forth, slowly and gently, like a baby in a cradle, is getting shorter, the force of the pull is getting stronger, unbearable. All of these because the question lingers, echoes, with a growing intensity: *what of the present in hums?* Am I locking them away? Like Perseus holding the monster's head? A myth is a myth, it can go only so far. Another quality of a myth, the reason why it fascinates us, Calvino points out, is *quickness*. To vouch for its value, he tells a story of a certain emperor, Charlemagne, who fell in love with a girl. When she died, he embalmed her body and continued his love with it. This is certainly a morbid event, full of potential gossip. Worried, the Archbishop examines her body, finding a magic ring. But once it is in his hand, the emperor's love moves to him. Embarrassed, he throws it into a lake, a place that the emperor since refused to leave. Calvino then points out that the reason this tale is fascinating is due to the speed in which a series of self-contained events are sewed up, and it is the magic ring that establishes the continuity and charm.[60] But there are so many things we do not know. We don't know why the ring appears, and how come it has that power. We don't know where it came from, nor do we have a clue how the girl got hold of it. Similarly, we don't know what really happened to Medusa's head, whether Perseus disposed of it, how he did it if he did, and how her head got that dreadful power in the first place. We know none of these details because the head and the ring are there not to oscillate between presence effects and meaning effects; they are present to create that force of oscillation. A sequence of events run and keep on going with lightness and quickness, and we are left, co-oscillating, with the effects of the head and the ring, be they present or taken as a meaning. But when we cannot help but face the head or the ring, the very force of oscillation, what do we do? Looking away or glancing sideways at a mirror would not work for hums that stay and, keep on staying, neither from a past nor to a future, but always at a present moment. Those are hums of the other.

3

Hums of the Other

Invitations to hums that have ears

As John Cage keeps on talking in *Lecture on Nothing*,[1] and again in *Lecture on Something*,[2] he does not stop and ask people to 'listen!' He knows very well that the audiences will have no choice but to listen; that they will have no control of what is being said to them. They know that sometimes, the talking is not only just for them, but that in those 'sometimes,' Cage is in it, too; both Cage and the audiences are out of control of what is being heard. In *Invisible Cities*, Calvino borrows the mouth of Marco Polo to point at that which makes audiences, and Cage, run out of control:

> 'I speak and speak,' Marco says, 'but the listener retains only the words he is expecting. The description of the world to which you lend a benevolent ear is one thing; the description that will go the rounds of the groups of stevedores and gondoliers on the street outside my house the day of my return is another; and yet another, that which I might dictate late in life, if I were taken prisoner by Genoese pirates and put in irons in the same cell with a writer of adventure stories. It is not the voice that commands the story: it is the ear.[3]

There it is, *the ear*. Contrary to what we have been led to believe, it is not the mouth, but the ear that is voracious, always seeking attention and care, and stretching us out. We do not and cannot eat all the time, but we get to hear everything, anything, all the time even if we do not want to (and I believe that even the most severely deaf people would have a desire to 'listen,' even if not to sounds;

they would desire to 'listen' to vibration, sensing, proprioception, stirrings, anticipation, and so on).[4] As the mouth does all it can to make us visible, speak, heard, alive, love, hate, complain, harass, offend, defend, and so on, the ear remains; it stays calm and keeps in the mode of collecting, reaching out, accumulating those which come to it; it keeps on hearing, even when we choose not to hear. This remaining-inside and keeping-at-hearing silently gives the ear a sense of control, and when its skills are harnessed, it even offers power. It keeps our world in check. Even though it is often accused of being gullible, the ear is still safe until the mouth admits it. The mouth, on the other hand, plays havoc with the calm world, plots schemes, and then spills out their secrets to the world. Perhaps it is in this reason why we have had a long, complicated, and contentious relationship with the voice: while the mouth spits it out and regrets, the ear collects it back and judges; in doing so, the self is divided, tormented. Therefore, if the self is in doubt, on the verge of collapsing, on the edge, just as he or she faces the other, in one of its various shades, it is not through the mouth, but through the ear, and when the mouth confesses it.

How can I speak of the hums of the other, those which do not stretch from me or my mouth, but come to me and to my ears, uninvited, unannounced, assuming me and my voice, whose ontology springs from silence and its process of negation? How can I speak of those of the other, of which mere presence would mean the stretching of the hearing, listening, performing subject out and into its void? Is it not that this question is in fact putting the mouth in a position of confession: to have my mouth say, I, or rather, my *ears* have encountered, a hum? It is in this sense that we must start our investigation of the hums of the other, not from the mouth, not from the voice, but from the ear. We will have to revisit the voice, as it connects the mouth to the ear, but at this time, a time in which we confess our encounter with hums of the other, we must do so through the ear.

We will keep returning to this question, as it is our ultimate aim for this chapter. For now though, we will turn that question into something less grave, more approachable. Let us return to Calvino and one of the virtues he held dear, *lightness*, because one aspect in his story – the relationship between Medusa and Perseus – is going to help us do that. In *Metamorphoses*, Calvino continues, Perseus wants to wash his hands after killing another monster, and Calvino is keen to observe how Perseus deals with Medusa's head. Interestingly, the hero performs the act of placing the head on the ground with care and gentleness: he makes the ground soft with leaves and branches of water plants, on which he places Medusa's head, face down. From this Calvino identifies not only the lightness of his gesture, but also the clashing image of the head that is 'so monstrous and terrifying yet at the same time somehow fragile and perishable'.[5] *The monstrous and terrifying, and at the same time, the fragile and perishable* – I do not know what else could more aptly describe what we are aiming at in this chapter. It is with such an understanding that I will introduce another story by Calvino, *A King Listens*.

> Sunk on your throne, you raise your hand to your ear, you shift the draperies of the baldaquin so that they will not muffle the slightest murmur, the faintest echo. For you the days are a succession of sounds, some distinct, some almost imperceptible; you have learned to distinguish them, to evaluate their provenance and their distance; you know their order, you know how long the pauses last; you are already awaiting every resonance or creak or click that is about to reach your tympanum; you anticipate it in your imagination; if it is late in being produced, you grow impatient. Your anxiety is not allayed until the thread of hearing is knotted again, until the weft of thoroughly familiar sounds is mended at the place where a gap seemed to have opened.[6]

To me, the most curious word here is *gap*. In *A King Listens*, the king initially believes he is in control; the way he controls is by listening. As such, his palace is made with 'all whorls, lobes: it is a great ear.' The king sits on the throne, 'in the innermost zone of the palace-ear,' and thus, his palace 'is the ear of the king.' The pride of this control over his kingdom is exuberant in the way the king conducts his listening. He has only 'to prick up' his ears to 'recognise the sounds of the palace.' And the sounds that he hears are the safeguards that help him to make sure that everything is in control – until he hears a gap. A gap, here, appears as a lack, an opening that needs to be filled up, as he, as a king, cannot allow such an opening. In the midst of creaks and clicks, flutters and clanks, drips and drops, pops and taps, peeps and snaps, shrieks and hisses, snaps and rips, squeaks and tweets – in the full, thick, surface of those sounds that reverberate, through which the king is able to survey his palace, a gap means a silence; a silence means a warning, as 'the threat comes more from the silence than from the sounds'.[7] A silence is an opening of his palace, which has otherwise been tightly guarded. And the mechanism of this guarding, this defence is through identifying his sounds, not just knowing what they are, but knowing their meaning, their signification, and significance: 'Alert, intent, you intercept them and decipher them'.[8] Even his spies, secret agents, embedded in every corner of his palace and kingdom send reports to him, but he does not need to read them ('No one, for that matter, thinks you must read the report delivered to you'). This is because, 'to be assured, you have only to hear the clicking of the electronic machines coming from the secret services' offices during the eight hours established by the schedule';[9] his ear is in control of his safety, his status, his palace and his kingdom. His alertness and a great scheme of listening notwithstanding, the king grows suspicious. Of what? Someone betraying him and dethroning him, of course, he is a king. In the story, the beginning of his suspicion – though we know *that symptom* had begun long before, immediately after he had dethroned the previous king and took his office – starts with this:

> Here the walls have ears.[10]

This is merely a saying, an analogy, initially, or just an image. But it is also the start *and end* of suspicion, or symptom. To the king – who believes in the power of hearing and who takes pride in the control that his ear exerts on those who are to be heard – is to admit a crack of his sovereign power; it is to confess that, perhaps, it is not just him who hears things. What arises alongside this suspicion is equally perturbing: who else is hearing what I hear? And if you want to push it further: what else is hearing what I hear? Still further: am I hearing that which is hearing? If you really stretch it towards a terror: *am I only hearing things that which is hearing?* We shall remember this question, to which we will return soon.

An equally horrific story that connects us again to the ear – this time with wit – is *The Metamorphosis* by Kafka. Gregor Samsa, an office worker whose life we can assume is mundane, wakes up one morning to discover he has become a gigantic insect. Strangely, and also rather ironically, he does not find this at all terrifying – until he hears himself trying to answer his mother, his voice now transformed:

> That gentle voice! Gregor had a shock as he heard his own voice answering hers, unmistakably his own voice, it was true, but with a persistent horrible twittering squeak behind it like an undertone, which left the words in their clear shape only for the first moment and then rose up reverberating around them to destroy their sense, so that one could not be sure one had heard them rightly.[11]

Gregor's horror has less to do with the sound of his voice and the persistent and twittering squeak that issues alongside it. Rather, his dismay has more to do with the fact that he has heard this alien sound with his own ear, reverberating through his own body. Gregor recognizes the voice as his, and yet, it is not his own volition that creates it. And thus, it is not the mouth, which emits those squeaks, that plays havoc, but rather the ear, which hears them, that throws him into horror. And it is the ear, not the voice, which questions the genuineness of that voice-squeak. Moreover, the shock and the following horror of hearing this voice arise from not knowing how the voice has come about; that is, Gregor has no idea how the voice was made. Hidden from its production, he is helpless and out of control under the effect of the voice. One peculiar aspect about Gregor's voice is that for the first few times when he tried to speak, the sound he heard was still a voice in the process of becoming that which is not, or a voice of a becoming-animal. In terms of the voice, it can be taken as the last trace of Gregor as a human being. And this in-between voice – the voice of becoming-insect, a voice that his father, his mother, and his sister could still make out – was in fact hidden from the rest of the family, behind the door. But then, we are not sure if the family really understands Gregor's becoming-insect voice at all, or if they just assume some kind of meaning for his squeaks. When he finally becomes fully visible to his family, they can no longer decipher his words; the becoming-insect voice no longer has any trace of the human. In any case, Gregor speaks little now, other than hissing or speaking to himself, which he does without any problem, fluently and satisfactorily.

Kafka makes use of becoming-animal voices in many of his stories to designate his hero or heroine as an animal, but never to borrow the archetypal form, signification or significance from that animal by doing so. Rather, he uses it to materialize each voice as a 'movement,' 'threshold,' or 'vibration,' and to realize an escape or exit for the character. Thus, Deleuze and Guattari point out:

> In the becoming-mouse, it is a whistling that pulls the music and the meaning from the words. In the becoming-ape, it is a coughing that 'sound[s] dangerous but mean[s] nothing' (to become a tuberculoid ape). In the becoming-insect, it is a mournful whining that carries along the voice and blurs the resonance of words. Gregor becomes a cockroach not to flee his father but rather to find an escape where his father didn't know to find one . . . to reach that region where the voice no longer does anything but hum.[12]

Among these becoming-animal voices, however, Gregor's insect-voice is a better example of the presence effects of the hum.[13] Firstly, it exhibits the terror of hearing a voice that is both one's own and something completely *other*. Secondly, Gregor's insect-voice never carries meanings or significations, and there is no potentiality or possibility that it will do so in other contexts or situations. Thirdly, this means that Gregor's voice persists completely outside of any networks of signifying chains. The family never understands Gregor's words, and his movements and behaviour are wholly mistaken. Fourthly, we have never, really, got to witness the process of Gregor's transformation; rather, the metamorphosis has happened, and he has become a gigantic insect. In this way, it is not Gregor's voice that has become that of an insect, but that becoming-insect voice has been installed on his throat, his mouth over the night. The metamorphosis, the process of becoming, thus, is directed not at his body or voice, but at the relationship that Gregor has with his body and voice, with his room, with his family. Finally, Gregor's insect-voice quickly alienates anyone who hears it. Gregor finally lets out a cry, making it possible for the visiting Chief Clerk to comprehend what is happening. He responds: 'That was no human voice.' Certainly, the family and the clerk are worried, caught in horror when they finally see the gigantic insect that Gregor has become. However, it is not his voice that has terrified them because they do not hear it *in the same way Gregor hears it* – simply speaking, they do not hear the horror, but only see it. It is only Gregor who hears the horror of his voice. This alienation of being subject to the symptom alone is a characteristic unique to the hums of the other. Even when such hums are experienced together as a collective, symptoms like anxiety and terror are solely for individuals and cannot be shared.

Not all those that appear in Kafka's world of becoming-animal whistle, squeak, cough, sing, or whine. There is one particular 'I' who says nothing but only listens: it is the 'I' in *The Burrow*. Like the king in *A King Listens*, 'I' in this story 'live[s] in peace in the inmost chamber of my house'. Just like the king, the main concern for 'I' comes from a suspicion that 'the enemy may be burrowing his way slowly and stealthily straight toward me'. Two activities content 'I': the first is taking pride in its burrow, the construction of which has consumed most of its life and that 'I' still spends most of the days revising and improving. The second activity is listening to silence in order to affirm stillness; thereby assuring itself that there is no danger. In fact, 'I' takes profound pleasure in listening; listening 'into the stillness which reigns here unchanged day and night'. But 'I' is aware that this silence is 'deceptive' as '[a]t any moment it may be shattered and then all will be over'. 'I' waits for danger – a danger whose reality is yet to materialize. Thus, 'I' spends as much time taking pride in its burrow as it does in imagining this materialisation of danger. This imagining is neither preparation for a counterattack nor a plan for dodging it. 'I' is well aware that any scheme such as this is futile. Rather, it is a *pure pre*-living of the danger that is to come.

Like the king's palace, the burrow of 'I' is symptomatic of what constitute the ear and its anxieties. For the king in *A King Listens* and the 'I' in *The Burrow*, the ear is both probing and controlling, but at the same time, the ear is a probe thwarted and its control foiled. It is helpless – and it knows! In what is coming, the ear's anxieties start. 'I' is almost exuberant when the danger that 'I' has been dreading finally materializes itself, showing itself to 'I' as 'an almost inaudible whistling noise' that awakes 'I'. No sooner does 'I' recognize the noise, than its ear starts probing, identifying the sound immediately as 'the small fry, whom I had allowed far too much latitude, [which] had burrowed a new channel somewhere during [I's] absence'. The probing ear is quickly thwarted by the realisation that 'I' cannot ever get closer to the source of the noise; it seems to be coming from everywhere and 'goes on always on the same thin note, with regular pauses, now a sort of whistling, but again like a kind of piping'. The whereabouts of 'I' notwithstanding, further investigation reveals that the noise persists. The awareness of 'I' – '[s]trange, the same noise there too' – is not only of the noise being hidden from 'I', but of the noise being everywhere. What can be both hidden from view and be present everywhere?

Besides this ubiquity, the everywhere-ness, however, the fundamental basis of the ear's anxieties, in whatever form they may take, is that this coming (which is devoid of those which constitute the King and his palace, 'I' and its burrow) effaces the temporality of its own coming – it is coming, to come, and to have come all at the same time:

> This noise, however, is a comparatively innocent one; I did not hear it at all when I first arrived, although it must certainly have been there; I must first feel quite at home before I could hear it; it is, so to speak, audible only to the ear of the householder.[14]

This is the chilling realisation of 'I', who 'suddenly begin[s] to hear now a thing that I have never heard before though it was always there'. In this sense, the omnipresence of the noise is not only everywhere-ness, but *everywhen-ness*; the anxieties of facing it are thus not only about where, but why now, why then or why before, as well as why not then and why not now. This is how hums of the other operate. They are pervasive. They are always already here, having-been-heard-of. You are always listening to them, now, here and there. They are everywhere, and most of the time, you do not feel the need to silence them, to remove them or to run away from them. As 'I' consoles itself, '[i]t is really nothing to worry about; sometimes I think that nobody but myself would hear it; it is true, I hear it now more and more distinctly, for my ear has grown keener through practice'. This is so for hums are cunning in the way they appear. The persistence of hums, the way they keep on humming, disarms whatever defensive scheme one might come up with. It is always too late when you realize that you have been trapped by them for they have already saturated you and your world, your senses, your hearing and listening; you are even participating in them, promoting them, both for the sake of, and towards the destruction of, your being. How do you get to become an accomplice in the hums? Your role as a co-conspirator has already begun by rendering your ear. In *A King Listens* and *The Burrow*, it is insinuated that both the king and 'I' may be perhaps the maker of the noise that they have been suffering from, and this hypothesis is only possible if the hums are not bound by time and place. And even both may be sharing their time and place: do you remember how the king's suspicion took shape? He says the walls in his palace have ears.

And we see that out of desperation, 'I' listens 'at the walls of the Castle Keep'.

But the presence effect of the hum *par excellence* is the buzz from a telephone. In Kafka's novel *The Castle,* K. phones in to find more about his assignment. K., not knowing exactly why he was summoned up by the Castle, is eager to have access to the Castle early in the morning on the next day, only to be told by his two assistants, both of whom he names Arthur, that a stranger to the Castle would need a permit to get in. So K. goes to the telephone and picks up the receiver:

> The receiver gave out a buzz of a kind that K. had never before heard on a telephone. It was like the hum of countless children's voices – but yet not a hum, the echo rather of voices singing at an infinite distance – blended by sheer impossibility into one high but resonant sound that vibrated on the ear as if it were trying to penetrate beyond mere hearing. K. listened without attempt to telephone, leaning his left arm on the telephone shelf. He did not know how long he had stood there, but he stood until the landlord pulled at his coat saying that a message had come to speak with him.[15]

This is a hum of the other calling, calling for K., just as K.'s ear is calling for it. Facing a hum of the other and listening to it does not need an attempt. There is no intention required, this is a sheer drive, a drive that pushes and pulls K. back and forth within the force of oscillating between what things mean and what they do not. This is the only explanation one can come up with for the phenomenon K. is experiencing: the sound that he has never heard before, which sounds like a hum but not just one hum, the echo of voices, singing, resonating, vibrating, penetrating. K.'s intuition guards him against the force by plunging both him and the hum into the signifying chain of the symbolic register that swings between meaning effects and presence effects. But it is futile, useless; it is on the verge of shattering, exploding into the thick, heavy air of nothingness. Both Dolar and Toop briefly discuss K.'s momentous, yet terrifying ordeal in front of the telephone, but through the framework of the voice. What K. hears on the line is '[j]ust a voice which is some kind of singing, or buzz, or murmur, the voice in general, the voice without qualification,' says Dolar. 'There is no message, but the voice is enough to stupefy him; he is suddenly paralysed'.[16] But this is outside of the voice; yes, the receiver on the telephone assumes whatever comes out of it to be a voice; we expect to hear a voice. In this regard, this is a voice, but *yet, it is not a voice*. If by 'the voice without qualification' Dolar means the voice *objet petit a*, a container, a void, a nothing, this hum – *yet not a hum* – has already surpassed it. 'If there is the voice, it is due to the fact that the signifier revolves around the unspeakable object,' says Miller, and 'the voice as such emerges each time the signifier breaks down, and rejoins this object in horror'.[17] Yet again, this hum starts neither from a signifier, nor the unspeakable object, each of which comes later by K. Due to the impotent attempts by K., there is no thread, no link, whatsoever; there is only a pure calling. And K. answers it by saying nothing to the call – he has nothing to say – instead he remains by being silent, hearing and listening to it. For how long he remains silent K. does not know, a silence that is soon broken in the next sentence: '"Go away!" yelled K. in an access of rage, perhaps into the mouthpiece, for someone immediate answered from the other end'. My reading of this is that someone answered K. as soon as he picked up the telephone, but K. lost track of time as he responded to the hum. Spellbound, K. perhaps did not want to leave the hum, and thus, 'Go away!' may have been directed not to the hum, but to the landlord who pulled at his coat or even to the person on the other end of the phone. This is a stretch of the hum, in a sense of the voice stretching the self. Here, the hum of the other stretches K. in such a way that he no longer counts time; he is displaced either to another arbitrary point in time, or to no point in time at all.

Humming and mysticism

As I have discussed above, Calvino's and Kafka's stories show that the hums of the other come to us in such a way that they are initially taken to be puzzling, inexplicable, and mysterious. The mysticism surrounding hums can, on the one hand, be ascribed to the gradual removal of the performability of humming from individual subjects or even from collective society. Just like the King in *A King Listens*, 'I' in *The Burrow*, or K. in *The Castle*, they can no longer participate in the production of the hum; what's left for them to do is only listen. The loss of ownership from the hum, set against the perseverance of the subject's desire to participate in sounding it, creates anguish. This anguish arises from his or her persistence in the hum, a persistence to keep on living, but at the same time, to endanger him or herself to that which no longer exists. LaBelle alludes to the anxieties of the ear when he posits the uneasiness of listening. Quoting David Michael Levin, LaBelle points out that 'the attention one may bring forward in that instant of listening, in *lending* an ear, is defined as "listening to the other"' to offer a 'mutual space'. LaBelle also points out that this listening

> is tensed by not only empathy and care, but also nervousness and hesitation, with longing and capture, by the invitations as well as the demands of the other – a combination that is highly suggestive for a culture of radical ethics and inclusion.[18]

This 'nervousness' and 'hesitation' are caused by the realisation that to listen is 'to give the body over, for a distribution of agency'.[19] Such a distribution of agency also means that one does not only have no control of what one can hear, but one also admits that impotence: 'Here the walls have ears'. The assumed control of one's participation in hums turns out to be a false hope in which we are powerless against and have no say in what we hear. What's worse, as the hums keep on humming no matter what they do, captive in listening, the experience of listening to hums drives the hum to and beyond the liminal state. At this point, you do not and cannot hear the hums any more, but instead listen to them intensely and forcefully; in fact, the allure of hums has become too strong for them to be background noises; you become K. In *The Castle*; or Alex in Stanley Kubrick's *A Clockwork Orange*, who volunteered to be strapped to a chair and undergo a series of psychological conditioning experiments. However, it would not be our eyes propped open, as Alex's were with some antique, mystical, mechanical device, but it would be our ears, propped open, always ready to hear. They would not be kept open by some device, but rather by our desire to listen and the drive to persist in listening, just like Alex's desire to watch. These are thus only a few rare moments in which the subject's desire and drive converge to bring about the devastating effect of cohesion. When you are situated on the side of the mundane, but are allowed to peek out, hearing through the liminal, you may say you are experiencing the Sublime, be it a failed representation of Thing-in-itself because of the very unattainability of the Thing, according to Kant, or the same description without its transcendent presupposition given by Kant, according to Hegel.[20] When you are pushed over to the other side, which you have been permitted only to glance at, peek through and hear sideways before – and it is never your choice and willingness to go over; you are pulled and pushed over – you do not become part of the Sublime. Nor does it become you. In fact, it is not the Sublime at all; even the Sublime has been subsumed by it. It is the Other in full power, the big Other, 'the Real in its utter, meaningless idiocy,'[21] just as the hum subsumes not only K. and his perceptual-signifying-survival system, but also the buzz (the signifier attached helplessly to the Silence by K.) that resonates the telephone, Kafka, his story about the Castle and K., and along with it, even us who read the story. It keeps on persisting while you are no longer; you are in the process of the 'subjective destitution,' defined by Lacan as the final stage of the psychoanalytic process, in which '*the subject no longer presupposes himself as subject,*' that is to say, he '*annuls himself as subject.*'[22] This is the primal terror of the hum: *the annihilation of the subject subsumed by his or her drive to listening.*

The terror of the hum and its all-encompassing force to *delete* us can also be traced to a rather unexpected condition: the relationship between the subject and his or her mother. Toop discusses a book by Anna Karpf, *The Human Voice*, in which she discusses how unborn infants identify their mother's voice, heartbeats, breathing, muscular tensions and movements of the diaphragm to take comfort. Karpf notes that this 'emotional, auditory and tactile bond' continues for four years until this integrated system is gradually substituted with a 'growing predominance of the visual' over the other senses. This is accessed, she argues, via

> these auditory origins, in the manner through which the masked echo of the maternal voice or its simulacrum may be a presence of problematic diversity and subterfuge through childhood into adulthood, even to death, or through mysterious preferences for noise that is filtered, immersive, physically overwhelming, rhythmic or in some other way, via distant memory, intra-uterine.[23]

Are these not the first hums, the sounds that unborn infants hear, those which are muffled, coming from everywhere, along with the tactile – its whole body feels the sounds! – a care of being held, encased in a thick, dark, liquid space? In and through these first hums, the foetus must have felt both secured and helpless. There is nothing it can do, no sense of control, let alone making its own hums, apart from persisting to survive, tied to the umbilical cord, in the fullest sense of contact with the mother's womb, and listening to those hums, hums that keep on humming. There is no meaning, no signification, no commitment, no guarantee to persist, attached to those hums: only pure acoustic saturation to no end. The 'huge psychological implications', as Karpf argues, 'linked to the intimate connection between maternal voice and secure attachment'[24] can be, on the one hand, those of the primal care and love, and on the other hand, those which could be the most insidious and self-denying. Not only the sense of feeling secured and attached, the first hums from the womb also bring with them that of impotence: the beginning of one's life, at the same time, insinuates its end. The first hums whisper the peril of the subject's being-in-the-world. In this sense, the image of a mother's hums to her baby – one of the most archaic images of unconditional love and dedication – may imply a much darker, sinister desire for control, power, and dominance. No wonder it is often used in film and mass media to portray just that.

On the other hand, the mysticism attached to hums of the other may also be explained by focusing on the phenomena experienced by 'I' in *The Burrow* and K. in *The Castle*. Taken hostage by the hum that keeps on humming, both 'I' and K. feel as if they were completely out of time, or as if time sustained or jumped to arbitrary points. According to Ihde, such an *obliteration of the timefulness of sounds* may be due to the way we take note of a thing's temporality. Ihde contrasts a mute, stationary object, such as a calendar on the wall, with a moving ball: while the former 'stands out as motionless and mute', to which we detect 'only a massive *nowness*' without any dramatic coming-into-presence or passing-through, the latter, Ihde argues, 'allows a shift toward the *noemetic* appearance of successive time'. With sound, this temporality 'is not a matter of "subjectivity" but a matter of the way the phenomenon presents itself,' says Ihde. 'I cannot "fix" the note nor make it "come to stand" before me, and there is an *objectlike* recalcitrance to its "motion."' If, however, one is presented with 'a single, sounding tone which does not vary and in which the depth of foreground to background features is eliminated' like the hums that both 'I' and K. heard:

> this presence can not only be deeply disturbing, but it begins to approximate the solid 'nowness' of the stable visual object and time sensing 'returns' to its location 'in oneself'.[25]

What Ihde says is this: as opposed to *objectlike* sounds of which temporal span is with limit, and thus, allow us to employ our temporal focus easily, sounds that prolong, like hums that keep on humming, which do not vary and void its spatiality by rubbing out its relation to their contexts, fields, surroundings, end up rewriting our given noetic – intentional act of listening – relationship to sound and restructuring it much like we would do with a solid, *mute, motionless*, object that we see. And accordingly – and this is most crucial for hums of the other – just as we turn to our consciousness to keep track of the object's temporality, we return to the listening self to make an attempt to keep track of the hum's temporality because the hum offers none. But as Ihde points out, this returning of time sensing to the listening self is an approximation of what we do with a stationary visual object. What ends up taking place with the returning of time sensing to the self is the listening self confronting the voice of the other in the guise of the hum.

In large part, though, the sublimity of hums is associated with the oral poiesis and consequent oral imaginaries that the mouth concocts. Examining 'unmoored, mouthless sounds' in *Beyond Words,* Connor notes how the oral poiesis of these vocalic sounds, such as 'their quasi-vocality, and their link to that familiar of the mouth, the resonation of the nose, and all the forms of impersonal or sacred noise which nasals can consequently evoke,' help them 'retain their implicit reference to the mouth'.[26] We lose the mouth to the hums of the other that we do not and cannot participate in, but only listen to; particularly, we lose all the physicality of the mouth that produces its fascinating materialities. Still, we attach that fascination of the mouth's materialities to the hums of the other. By doing so, we do not gain the oral imaginaries that LaBelle has thoroughly examined; rather, any potential oral imaginaries with (not from) the hums of the other become contingent upon the most precarious attempts to regain our mouth. If, as Miller has shown us, the voice's emergence is caused by the revolution of the signifier around an unsayable object, it is indeed a miracle – a myth – how we somehow attach any shape of orality to the hums of the other as we are not certain whether a signifier or an unsayable object is present with the hums; what we can assume with more certainty is that there is a force of revolution around what is being heard. And this force is our pure drive of the ear to listen. The attempts to attach the oral imaginaries to the hums are most precarious because it is through, and only through our pure drive of the ear to hear that these imaginaries take shape. In turn, this shows evidence of how strong the drive of the ear can be.

The aphonic, the acousmatic and the air

'We listen first to things,' Ihde declares.[27] Noting the visualist perspectives in the phenomenologies of Husserl, Merleau-Ponty, and Heidegger, Ihde takes a decisive auditory turn to human experiences. Just as our experience is that of something, he argues that our natural and ordinary listening is first and foremost toward things. But what *things* do we listen to when we hear hums, particularly those that push us over the liminal and over the Sublime? Are hums themselves things? In other words, can we talk about a hum as an independent, self-organizing element of whose characteristics we can then explore? Or is a hum something that constitutes or reconstitutes other things, which then should be our ultimate aim to examine? Can we get to hums themselves or should we let things show themselves through hums? All these questions are absurd; hums are of course not things; there is no *thing-ness* in hums that we can listen to; that is what the king in *A King Listens*, 'I' in *The Burrow*, and K. in *The Castle* have tried and failed; the hopeless attempts to find *things* in hums have led them to the powerful oscillation of signifiers after signifiers that are illusory. So we ask: where do these hums of the other come from, the ones that the king, 'I', and K. all heard and of which they fell under the spell? These are not things; there is nothing to hear, but at the same time, there are so many that they end up listening to.

Returning to Ihde, we see that he is keenly aware of this problem. '[W]ithout forgetting this first presence of the existentiality of the thing,' Ihde points out, 'the concern of phenomenology must also be expanded beyond any exclusive concern with things alone' because without bringing into question 'how things present themselves in terms of a situated context,' we will be falling into the pitfall of 'the illusion of a thing-in-itself'. Thus, the first declaration of Ihde can be more specifically stated like this: we listen first to things, not alone themselves, but 'within a field, a limited and bounded context'.[28] Ihde goes on to examine, slowly and carefully, what he means by this situated context, but we have already identified what it is with hums: hums are situated in the context of a tension between lack and surplus, and fundamentally, the lack-surplus tension is operated under the acute awareness of something missing, lost. In this sense, like the voice, hums, irrespective of those of the voice, of the voice-like, or of the other, are apt to be designated as a Freudian-Lacanian object since 'the object in psychoanalysis is always thought of as a *lost* object' and thus, being a lost object, hums become 'the object of the drive'.[29] In this way, hums can be taken to be an *object hum h*. What's more, we have also examined what is lost in hums and through this loss, and only through it, what is gained in hums. And three key features of hums among others arise through the workings of the bartering of loss-gain under the lack-surplus tension: *the aphonic, the acousmatic*, and *the air*, which I consider are responsible for pushing hums to and over the Sublime.

The aphonic

The dictionary definition of aphonic is 'having no voice or sound' or being 'mute'[30] due to the negative operation of *a-*, and is connected to the other two terms, aphasic and alingual, the former of which is an adjective of aphasia, 'inability (or impaired ability) to understand or produce speech, due to brain damage.'[31] However, the latter, alingual, is a more extreme form against what is deemed lingual, which means either 'relating to, near, or on the side of the tongue' or 'relating to speech or language'. Thus, the alingual could be understood as the inability to 'understand spoken, written, or gestural communication'.[32] On a more defined scope, the aphonic may mean 'not-phonic,' that is, those that are uttered by the mouth that are outside of the phonemic system. In this system, *aphonicity* is based on whether vocalic sounds have a membership to a language 'in the sense of a regular structure of similarities and differences'; that is, on whether they can 'embody meaning'.[33] Thus, contrary to its nominal definition, the aphonic, as we understand it, is not without sound; rather, as has been briefly pointed out through Miller and Dolar, it is the hollowing, empty, and void nature of hums, and thus, it is waiting for one, 'an unheard voice,' or a yet-to-be-heard voice, which will fill the gap, the lack. What is disturbing with this acknowledgment that hums are aphonic, though, is that we have no control of what will come to fill the lack. Recognizing that hums are aphonic is not a negation, but an invitation to whatever will come into being, which K. in *The Castle* experienced. It was not K. who invited the buzz-hum, but the buzz-hum invited itself. The conditions for the aphonic, as an initiation, have been noted: the mouth shut, the lips sealed, and thus, hiding the source of the sound, all of which will then lead us to what David Toop calls 'sinister resonance'. Hums 'have no reality as physical beings'. However, as Toop notes, 'just beyond reach', they are 'a deadly lure.' He continues:

> Sound is a present absence; silence is an absent present. Or perhaps the reverse is better: sound is an absent presence; silence is a present absence? In this sense, sound is a sinister resonance – an association with irrationality and inexplicability, that which we both desire and dread.[34]

On the other hand, the sinisterness of hums may come precisely from that hollowness, the emptiness of aphonicity of hums, which are devoid of any particular source to be attached, and thus, which invite any signifier. In this sense, the operation of the aphonic is tethered with that of the acousmatic.

The acousmatic

The acousmatic, which has been identified from the start of this book, is the ontological condition of hums: the mechanics of hum-making, i.e. a hum implies our mouth to be shut and lips sealed, to cause the complete modalities of the mouth to be hidden from view. As has been discussed earlier with regard to humming and acoustic mirroring, that is, how hums are initially bounced back by the closed lips and mouth and directed inwards, the modalities of humming are more complex than what the acousmatic as a concept would hope to highlight. The acousmatic has recently become a contested term. Initially introduced by Michel Chion in the context of *musique concrete* and the use of sound in film, the acousmatic is 'adjective, indicating a [sound] which is heard without the cause from which it originates being seen'.[35] As such, the acousmatic, also called the acousmatic situation, is nothing new; we hear sounds without seeing their causes all the time. Often, the acousmatic is used to connote a specific socio-cultural and artistic context that has been heavily institutionalized, such as acousmatic music, a terminology and practice mainly popular in academia in the UK and some part of Western Europe and America. Initially understood and implemented by Pierre Schaeffer and composers/theorists in his lineage, the acousmatic is tinged with phenomenological approaches, such as reduced listening similar to epoché and phenomenological reduction. Acousmatic practices were thus taken to be more of an attitude than a situation from which the term had purportedly originated; an attitude heavily conditioned by technological means that became available to Schaeffer. Recently, however, the institutionalized theory and practice of Schaefferian acousmaticism have been carefully examined and criticised. The most thorough investigation and critique on the acousmatic was conducted by Brian Kane with his book *Sound Unseen*. In it, Kane's objectives are broad in scope and comprehensive in methodology; first, he strives to tackle the mysticism surrounding the origin of the acousmatique or acousmate, as well as the oft-quoted Pythagoras' Veil, which has long been taken both as a method and as a (strangely distorted form of) power. Here Kane plays a myth-buster by conducting a thorough literature review and showing how such a story has been established as a myth. Furthermore, by taking such a critical stance toward Schaeffer's acousmatic and reduced listening, he also conveys a more serious criticism of phenomenology as being ahistorical.[36] His aim, however, is not to attack the acousmatic; on the contrary, his project works toward (re-)situating the acousmatic as a far-reaching theoretical and practical concept, a 'rubric intended to capture a set of historically situated strategies and techniques for listening to sounds unseen'.[37] A more focused study on Schaefferian phenomenology centred on the acousmatic and reduced listening was given by Kim. Kim takes a critical view that Shaeffer's implementation of phenomenology did not conduct a full sense of phenomenological reduction, and thus lead to limiting 'the totality of listening phenomena to its part, thus endangering the phenomenological project that it set out to do'.[38]

Despite, or perhaps due to, the (now highly contestable, thanks to Kane) myth of Pythagoras' Veil, the acousmatic voice is inherently visual, or at least, its effect arises from that which is visual, or more preciously *hidden* or *to be seen*. In this regard, the voice hidden is the voice with an intention or that which has been hidden by someone or something; in a similar, prophetic connotation, the voice to be seen is the voice that is assumed to be coming towards us and to be emerging; how soon, we do not know, but soon enough. And if this acousmatic voice that is currently not to be seen, is identified, assumed or imagined (which is more likely), this acousmatic situation turns into a powerful push-pull, and sometimes, dreadful game, as Toop has warned. Why are hums, acousmatic in nature, enigmatic and powerfully mystical? We can perhaps come up with three ways to answer this question.

Firstly, the voice in the symbolic register – following Lacan – set in motion a paradoxical game of something and nothing, presence and absence, surplus and lack. One consequence of this is that the voice's assumed attachment to the body is seriously questioned. Žižek notes that 'the moment we enter the symbolic order, an unbridgeable gap separates forever a human body from "its" voice,' which consequently gains power, 'a spectral autonomy.' He goes on to point out that this voice

> never quite belongs to the body we see, so that even when we see a living person talking, there is always some degree of ventriloquism at work: it is as if the speaker's own voice hollows him out and in a sense speaks 'by itself,' through him.[39]

And such a spectral autonomy, which is recognized by Connor as ventriloquism,[40] is fully realized with hums. The only contact that hums would have with the body would be its resonance within itself; other than that, there is no visible cause-effect relation between hums and the body; everything is completely hidden from view due to both the mouth and lips being silenced. On the one hand, while this autonomy causes hums to become deprived of the body, it also gives hums the air. What happens when the voice turns into a hum, through which it loses the body and gains the air, is this: the voice morphs from its inherent 'voice-body' into a single, pure vocalic entity that *materializes itself* – a true form of ventriloquism. This, Connor notes, is 'an idea – which can take the form of dream, fantasy, ideal, theological doctrine, or hallucination – of a surrogate or secondary body, a projection of a new way of having or being a body'.[41] On the other hand, turning into hums, the voice no longer needs to hold bodies and languages together, contrary to what Dolar noted.[42] Rather, being acousmatic, hums aim to leave off from the body and the mouth more precisely: having the quality of the air, hums can assume any body, or no body at all. At the same time, hums aim to be *alingual* – literally, neither related to nor towards the tongue; the tongue is dormant, set aside. And thus, hums are devoid of linguistic traits.

Secondly, once acousmatized from the birth, hums can never be disacousmatized. In other words, they can never be attached to a source, a body, an action; hums are fully acousmatic, and in fact, it is their ontology; hums are always somewhere, around, near, roughly, or about here or there; their production is always mystical, suspected, if not completely hidden. But in a sense, the voice has already had a difficult relation with its assumed body anyway, which Dolar discussed in more detail. First of all, he suggests, any body propped up as a candidate for the voice would never fit – even the voice's own body, a body where it is deemed to have originated from. There would always be a sense of clunkiness and ventriloquism, and the unveiling process of disacousmatisation – which reveals the source of sound and discloses the layers of what conceal the source from view, as explained by Chion – would eventually bring about neither its source nor a full unity of the voice and the body; it engenders a fetish:

> This is how Freud accounted for fetishism: one stops at the last-but-one stage, just before the void becomes apparent, thus turning this penultimate stage into a fetish, erecting it as a dam against castration, a rampart against the void. In this light we can grasp the whole problem of the fetishism of the voice, which fixes the object at the penultimate stage, just before confronting the impossible fissure from which it is supposed to emanate, the slit from which it allegedly originates, before being engulfed by it. The voice as a fetish object consolidates on the verge of the void.[43]

In a similar way, any attempt of disacousmatizing hums ends up encountering, or being thrown into, the void, engendering hums as a fetish, an object placed as the final layer before silence, towards which our desire is irrevocably driving. Dolar notes that the voice as the object, *objet petit a*, emerges in the midst of the 'impossibility of disacousmatisation', not through the impossibility of the voice being pinned down to a body. Interestingly, we arrive at the door of the same void, the voice of the other, not only by the failure of disacousmatisation expounded by Dolar, but also by the broken chain of signification, as shown by Miller.[44] But for hums, and I believe for the voice as well, this inability to be attached to any body and of not being fully disacousmatized brings about their spell and power. This impossibility reminds us of, and at the same time lets us confront, paradoxically, where the tension and the desire of disacousmatisation comes from.

Finally, by continuing to hum, hums aim to bring out those that do not sound – silence. These are not silences in the Cagean sense of the symbolic order, but the silence of the other, in the register of the Real. To explain this, let us posit a peculiar situation of the acousmatic, one that does not often come to the fore, which Dolor termed an *acousmatic silence*. Miller points out that Lacan's object voice did not come from a phenomenological description of the self's 'monologue in his solitude', but from clinical experience, from a psychoanalyst's observation of patients during sessions. In the process of analysis, it is the voice that guides the whole trajectory of the session. The voice is 'the very medium of analysis', 'the only tie between analyst and patient'. This silence, the silence of the analyst, exerts its power on and over the whole session, formulating and responding its voice without utterance. Dolar continues:

> The analyst is hidden, like Pythagoras, outside the patient's field of vision, adding another turn of the screw to Pythagoras' device: if with Pythagoras the lever was the acousmatic voice, then here we have *an acousmatic silence*, a silence whose source cannot be seen but which has to be supported by the presence of the analyst. All three modalities of Freud's voices are gathered together and come into play.[45]

Explaining what this acousmatic silence may signify, Dolar argues that this silence is that of the Other, which is in conversation with the patient's voice, now taking a special form of *lalangue*. This is '"the voice in the signifier", the coded message pertaining not to the speaker's intention but to its slip, the constant slipping of the signifier on the voice'.[46] Such a unique and rare incident in which the 'voice' – that is, the acousmatic silence of the Other – appears in its full sense, would have only been possible due to the peculiar context of a psychoanalytic session. Here, I recognize a similar silence, a far-fetched, at first, but gradually becoming indistinguishable, silence: the silence of the ontological voice of Dasein by Heidegger, of which Kane conducted a detailed examination. According to Heidegger, Kane argues, the ontological voice of Dasein is silent, without any 'specifically sonorous aspects'. And this is not just simply silent, but 'according to Heidegger's description, "a keeping silent"'.[47] Three things to note here, with this strange description: first, this *keeping silent* may signify a care for being silent; second, it may also point to a silence imposed upon threat; and at the same time, and it may denote the tenancy or persistency in being silent. In consideration of these potential meanings, we also need to posit the other aspect about silence of the ontological voice posited by Kane – a drive to disacousmatize the ontological voice:

> [A] sound becomes acousmatic when a listener apprehends the spacing between source, cause, and effect. Spacing is not an entity; it is nothing that can be found in terms of objectively present (i.e., present-at-hand) things. The strange phenomenon (if one can still use that word) of acousmatic sound surges forth only with the spacing of the source, the cause, and the effect. Dis-acousmatization occurs at the moment when the spacing of source, cause, and effect is overcome or banished by locating an object that can occupy the position of the source and reestablish its plenitude.[48]

Taking into account the desire of disacousmatisation in this way, we are getting nearer to the true nature of hums of the other. The desire to get closer to, remove and set aside the spacing between the source, cause, and effect of hums is non-negotiable. This is an attempt to get to Dasein, as well as Dolar's concept of disacousmatisation, a project that may ultimately prove to be foiled. Here, the secret is the word 'ultimately,' a word uttered with hesitation and reluctance, not with exuberance and enthralment.

The air

The air has been recognized as an essential matter of hums that causes them to be suspicious, on the one hand, and to shed the body, and thus to assume anything or nothing, on the other hand. One particular, certainly quirky, aspect of hums that the air may be highly responsible for would be that hums may connote smell. When I arrived in Aberdeen, Scotland in January 2013 to start my work at the University of Aberdeen, my project on humming, which had by then matured as an artistic, public sound art project, was slowly expanding to become an interdisciplinary project, part of which would consider the notion of humming as a sonic concept. It was around that time that a colleague of mine pointed out that the connotation of a hum or humming was not all that wonderful or charming because in the UK, a hum would mean a completely different thing. And surely it was: 'to hum', in British informal, is 'to smell unpleasant'. And there is another: hum is an exclamation of expressing 'hesitation or dissent', related apparently to 'hum and haw,' which means 'to be indecisive'.[49] What are these anomalies in the use of hum, these noises that trouble the map? The latter, being indecisive, can be ascribed to the mouth shut by the act of humming, which we have discussed; someone whose lips are closed and mouth shut is not ready to speak out, as he or she is indecisive. The former, however, is trickier, and its connotation, its image, more troublesome. What does humming have to do with smelling? I believe this has to do with the air, caused by the mouth closed; those whose mouth is closed as they hum end up breathing in and out with their nose, hence the smell. But this doesn't really solve the conundrum: why does that smell have to be *bad*, why such a negative connotation? Again, this – and 'to hum' or 'hum' connoting indecision and suspicion – has ultimately to do with the air, for there is something in the air.

In *The Matter of Air*, Steven Connor's fascinating exploration of the meanings of air as he follows the Western mind's inquisitiveness on the hazy matter in religion, psychoanalysis, science, art, literature and philosophy, Connor argues that for the last four centuries, the Western perspectives on air have been transformed so dramatically that it has now begun to be treated as a theme for scientific, technical and philosophical studies. We can literally occupy air, inhabit it, as we can extract and manufacture air for our own specific use. Still, the air is considered dubious, doubtful, and full of unknown intentions:

> Human beings have always believed themselves to be in part airy, and have often wanted to believe that their most essential part – their spirits, as they have liked to call them – were aeriform. And yet, in their clinging humility, human beings have viewed the idea of a literal translation into an airy condition with suspicion. *For airiness also signifies delusion, insignificance, even madness.* When Marx and Engels declared that, under capitalism, 'all that is solid melts into air', they were being prescient, but not approving.[50]

Connor goes on to examine various permutations of air, including haze, an element of atmospherics. In addition to so-called 'Romantic haze,' which refers to 'the haze of glamour, or diffused radiance,' modernism was endowed with another association with haze, a suspect and malicious matter which will endanger perception as well as bodily health, and by contaminating the pure, crystalline transparency of eye and mind,' haze will engender dreams and delusions.[51] The same goes with other types of gaseous vapour, such as 'fog, smog and other airborne emanations in urban environments,' at which the traditional hostility 'towards the corrupt and corrupting nether air' was directed.[52] Little sentimental or romantic glow was placed on fog, particularly severe in the UK and Europe in the nineteenth century, resulting in a range of fantasies and phantasmagoric images.[53] Connor notes:

> Nineteenth-century depictions of fog, the most extended and uncompromising of which is, of course, Dickens's at the beginning of *Bleak House*, inherit the vaporous sensibility of the medieval and late modern world, for whom mists and fogs are held to be the unhealthy halitosis of the ground, constituting a funerary air, full of infection, as opposed to the ethereal lucidity of the upper air.[54]

Perhaps it was this stench of the foul air that might have been pervasive in the daily life of the British people in the nineteenth century that caused the hum to assume this unexpectedly negative meaning. Whatever the case, it seems reasonable to posit that such a close link between hums and the air may have formulated the connotation of a hum being indecisive, suspicious or doubtful. In fact, the undertone of the air was so unfavourable that some chemists in the eighteenth century in France and Britain, the period which Connor termed the 'Chemical Revolution,' sought to replace air with more trustworthy and principled terms. Notably, Boyle, Black, Priestly and Davy in Britain employed variegated terms that denoted 'air' to separate from 'common air the elastic fluids they produced in their experiments.' Shown in *Method of Chemical Nomenclature*, which was first published in France in 1787 by four French chemists, and translated into English in the following year, are attempts to clear away dubious and confusing names 'inherited from medieval alchemy.' Connor lists some of the changes proposed in the book:

> Among the most important of the changes proposed by the 'chemical revolution' was the use of the word 'gas' in place of the terminology of 'aeriform fluids', 'elastic fluids' and 'airs' used by earlier investigators, and commonest among British chemists. So 'fixed air', also known as 'mephitic gas' and 'solid air', became 'carbonic acid gas' (later carbon dioxide); 'inflammable air' became 'hydrogen'; 'marine air' became 'muriatic acid gas' (later hydrogen chloride, which becomes hydrochloric acid in solution); 'vitiated air' became 'azotic gas' (later nitrogen); and, most importantly, 'vital air', or 'dephlogisticated air', became 'oxygen'.[55]

Even so, it must be noted that regardless of these dubious and even dismissive significations and connotations of the air by which a hum or humming may have been contaminated, the air constitutes humming, and it does so literally, figuratively, substantially and allegorically. What's more, the air encompasses humming in the sense that air is most essential matter for a hum to be actualized, but it is also essential for everything else. At the same time, the air overwhelms humming; no sooner does it constitute humming than it negates and de-constitutes it for 'nothing maintains itself in the same way any longer in air.'[56]

There are other significations that the air may bring to hums: firstly, possibility. If you remember, we started our journey into humming by drawing the air into our lungs. But we have soon realized that this possibility is not on and toward the rational and the systematic, but on and toward the mysterious, the suspicious and the magical. Connor notes the difficulty to stay balanced between 'two strains, of magical belief, and the sceptical anatomy of' vocalic sounds that continue to become outside the phonemic system.[57] For humming, though, the possibility of the air is always mysterious, suspicious, and magical, and thus, on and towards the magical or phantasmagoric potential. Another one, which is related to possibility, is the generative. The air taken into the lungs is the air used to generate sounds. When this is combined with the air's phantasmagoria, the aerial imaginaries, hums take flight to the air; from there, hums promote and conjure up the magical mouth and its magical mimesis to which we become obsessively drawn. Here, the oral imaginaries of LaBelle are commingled with aural imagination.

Many philosophers and scholars have posited the qualities of the air, and I will run a quick survey on those, particularly, by Connor, Bachelard, and Irigaray. First, Connor points out that the air was traditionally understood to be ubiquitous, yet plural. He argues that this is why the air has been less conceptualized. But, he notes, there are a few exceptions:

> Hebrew had no single, all-encompassing word for 'air', in all its aspects. In this it resembles Sumerian. Although there is a Sumerian god of the air, Enlil, there is no term for the air at rest in Sumerian, or in Akkadian, all the terms in fact denoting various kinds of motion – wind, storm, etc. . . . [T]here was in Greek thought a strong tendency to see the air as multiple rather than singular in form. Indeed, its conception of the air was, as it were, distributed between two words: the *aether*, the realm of brightness above the clouds, and *aer*, the dim, moist atmosphere below them.[58]

Thus, the air, for Connor, emerges as its form through its qualities or attributes, such as 'accidents, and appearances – as breath, wind, height, space, transparency, lightness, light'.[59] Furthermore, the air is not so easy to be contained for its nature is to be everywhere; which also means to be nowhere and always absent. Yet, despite its omnipresence and non-absence, air is difficult to isolate. This quality of air gives rise to a fundamental problem for science, a system of epistemology that asserts its power by way of identifying, dissecting and separating individual parts in order to understand the whole.

When air finally succumbed to scientific procedure, which Connor details in *The Matter of Air*, it 'had to be divided from itself, deprived of its principal features, namely its tendency to mix, compound and diffuse, along with its ubiquity'.[60] Paradoxically, though, this scientific process can never conquer the air in its full sense. For the air is a being (or becoming, perhaps) of negation, or an *unbeing*. As opposed to other elements that 'present themselves under many different forms', the air is extremely deceptive in its appearance, easily diffused into something else, and is always ready to disappear. By negating its own existence, Connor posits, the air negates 'perhaps even negativity itself':

> Take away the air, and the empty space you have left still seems to retain most of the qualities of air. The air is unique among the elements in having this affinity with nothingness, in signifying the being of non-being, the matter of the immaterial.[61]

In spite of being an unseeing due to its negation, though, the air does not exclude incongruous thoughts that may be paradoxical. It has room for the 'idea of the infinite, the illimitable, the transcendent' because of the air being regarded 'not as an element, or as a substance, but as a dimension'.[62]

On the other hand, Irigaray considers the air to signify that which goes beyond, that which transcends and exceeds any attempt at containment or identification. But 'this element, irreducibly constitutive of the whole, compels neither the faculty of perception nor that of knowledge to recognise it'.[63] For her, the virtue of the air is to be everywhere, always, and thus, allowing us to forget it. But this forgetting of the air is not that of the *materiality* of the air; rather, it is the forgetting of its attribute, or its operation of appearing and disappearing, being present and being absent.[64] Similarly, Bachelard takes the air to mean poetic dreams, escapes and enlargement. Throughout his book, by way of his writing which is noted for its poetic and fleeting beauty, he proposes that a study of air and poetic dreams is that of *'fleeting images*,' and '[i]mages of aerial imagination either evaporate or crystallise'.[65] He further notes that 'a psychology of the imagination,' which we have been employing for the examination of hums, 'cannot be developed using *static forms*'. Rather, it should be the other way around, that is to start from 'forms that are in the process of being deformed', focusing on the very 'process of dynamic principles of deformation'. The next paragraph by Bachelard is not only beautiful, but also pertinent not just to the aim of this book, but towards future questions and doubts raised by its discourse:

> If there is a dream that is capable of showing the *vectorial* nature of the psyche, it is certainly the *dream of flight*. The reason is based not so much on its imagined movement as on its inner substantial nature. Through its *substance*, in fact, the dream of flight is subject to the dialectics of lightness and heaviness. From this fact alone, dreams of flight can be divided into two different kinds: light flights and heavy flights. Around these two kinds are grouped all the dialectics of joy and sorrow, release and fatigue, activity and passivity, hope and regret, good and evil.[66]

Taken to be 'not a place, or a substance,' but 'a direction, an expansion, and multi-vectorial potential',[67] the air for Irigaray and Bachelard brings a lighter, and certainly more positive and less ominous, nature of hums than those of hums I have been examining. In a sense, this is what Bachelard meant by the dialectics of lightness and heaviness; commit yourself fully to those which are light or those which are heavy, even allow yourself to forget that those are either light or heavy, but never forget that you are in a process of dialectics.

In the end, the aphonic, the acousmatic, and the air of hums bring out not what hums are or have been, but how they will become. What I mean by this is that, just as Bachelard points out so gracefully, as soon as we take these characteristics to depict the various layers of the static features of hums, we detach from the dialectics of hums that are ultimately a dream. Being a dream, a dream of flight, hums follow the structure, the narrative and the logic of a language unknown and never-to-be-known, to us. Connor, surveying vocalic sounds beyond words, shares a similar difficulty, admitting that 'the magic of sound iconicity and its associated forms of mouth-mysticism are ultimately founded on – nothing; founded, that is, on the fact that the source of this magical power is just noises made by air, cartilage and saliva'. He goes on to posit:

> The mouth is magical because, as the vehicle of speech, it constantly translates hardware into software, making something

(meaning) out of nothing (mere sounds), and nothing (idea) out of something (matter).[68]

And from the heavy (and light) traffic of this transcoding between something and nothing and back to something, it is imperative to identify which oral imaginaries of the mouth we are traveling through.

Hums and oral imaginaries

Hums embed in their aural substructure the image of sealed lips, which, upon careful examination, unveils fascinating shades – secrets – of humming. The first, direct connotation of the lips sealed is of being mute, which in turn may signify many other things. These may vary from a willingness to stay silent in order to listen, to a vow of silence invoked as part of religious practice. It may include the alingual, a course by which the subject is silent due to little or no knowledge in a language. And it may also suggest the aphonic, which is ascribable to aphonia or aphasia, both of which are the condition in which one loses control of one's voice, either due to physiological damages to the throat, larynx or mouth, or caused by damage to the brain. Muteness also connotes ventriloquism, fasting, the death-drive, silence, and the simple line that the lips create when they are closed.

Dolar points out that according to Freud, all drives are 'mute, silent'. However, this silence is not just 'stillness, peace, an absence of sounds'; rather,

> in its proper sense it is the other of speech, not just of sound, it is inscribed inside the register of speech where it delineates a certain stance, an attitude – even more, an act. For silence to emerge, it is not enough just not to make any noise, and the act of the analyst hinges very much on the nature of silence.[69]

We have already examined this 'silence' in great detail in the previous chapter – how it comes about and how it gains momentum – through the silence of Cage. And later in this chapter, we will return to this silence again as all facets of hums of the other, from oral imaginaries to key features like the acousmatic, the aphonic, and the air and its various symptoms, some of which have already been examined in the stories of Kafka and Calvino. Here we are more interested in the oral imaginaries that the sealed lips would connote, and the first image of the sealed lips is a vow of silence.

A vow of silence in Christianity is a way for the subject to demonstrate appreciation for the transcendent existence of God. According to Sells, one can posit three approaches to face the 'primary dilemma of transcendence', which is to say, how to announce the divine being who is transcendent. The first is by utilizing different kinds of naming for the transcendent based on ways in which it is beyond or not – for example, God or a god. The second is to accept the dilemma as unresolvable, which leads to negative theology, in the language of which is an interaction between *apophysis* (negation or *apo phasis*, which means un-saying or speaking-away) and *kataphasis* (affirmation, saying, speaking-with). Sells explains:

> Any saying (even a negative saying) demands a correcting proposition, an unsaying. But that correcting proposition which unsays the previous proposition is in in itself a 'saying' that must be 'unsaid' in turn. It is in the tension between the two propositions that the discourse becomes meaningful. That tension is momentary. It must be continually re-earned by ever new linguistic acts of unsaying.[70]

The third is silence, as explained by Augustine, a solemn confession that there is no word worthy of God that can be uttered or announced. In this case, it would be better to say nothing than to try to speak about Him. How does he know this? For God is ineffable:

> If what I said were ineffable, it would not be said. And for this reason God should not be said to be ineffable, for when this is said something is said. And a contradiction in terms is created, since if that is ineffable which cannot be spoken then that is not ineffable which can be called ineffable. This contradiction is to be passed over in silence rather than resolved verbally.[71]

A vow of silence is in line with this belief, a belief here I mean not in God, but in the ineffable, a belief in the force, the tension, of being mute. What else would be more apt examples of such imagery than the films by Tarkovsky? The force exerted on the subject by being muted in Tarkovsky's films are key attributes of the main characters he portrays. Here, the condition of being mute is explored and intertwined in the midst of all possibilities and constituents that are both internal – the story, the characters, the form and structure of the film and so on – and external – the solemn confession of the pious director, his difficult relationships with the Soviet regime, the consequent long exile from his country, longing for his homeland, family and root, and for his fight with terminal cancer. In his films, Tarkovsky connects the condition of being mute to his heroes' commitment to speak nothing out of their religious or spiritual vow. Of particular interest to me is the vow of silence made by Andrei in *Andrei Reblev*, a historical drama based on the life of a fifteenth-century Russian icon painter. After witnessing the true cruelty of humanity through the Tartars' Raid in 1408, but also through his own, as he kills a man trying to protect a girl named Durochka (who is portrayed as a Holy Fool in Eastern Christianity), Andrei meets and converses with the now dead Theophanes the Greek, who tells Andrei, 'God will forgive, but you must not forgive yourself [. . .] Go on living between His forgiveness and your own torment'. In response, Andrei says, 'I'll take a vow of silence before God, and I won't say a word. I have nothing to talk with people about any more'.[72] Taken in this way, Andrei's vow of silence is less to do with the complete relinquishment of his self to God than to do with his resentment toward the sin of both humanity and himself. And perhaps his resentment is even directed towards God, as his actions are borne out of his deep antipathy to care and love. His action is thus in suspicion of his own mouth and voice as well as that of the human race; speaking would do no good. No longer speaking or painting, he enlists himself, after having returned to his monastery, to heat water by moving hot stones from a fire with tongs, an act that will remind us of the scene of Isaiah in the previous chapter where an angel carried a burning coal and 'cleaned' the lips of the prophet. Andrei's self-inflicted punishment would continue for sixteen years until he witnesses a miracle centred on the son of a bell-maker who had taken up an impossible task of casting a bell for the Grand Prince. Against all odds, and despite the bell maker's son not having any secret knowledge of how to make a bell, the bell is cast and produces perfect sounds. While comforting him and his tormented soul, Andrei finally breaks his vow of silence and tells him: 'You'll cast bells. I'll paint icons'. It is a mesmerising, and certainly symbolic, moment for him to speak again with the backdrop of the resounding bell.

A vow of silence, as the sealed lips and the shut mouth connote, is borne out of one's commitment to not open the mouth, or to speak. More fundamentally, while it is a gentle act, sealing the lips still acts as an oath to disengage one's mouth (and in fact, one's whole body – and soul) from the world. In this regard, it is an act that takes the mouth decisively away from its 'profound contact with the materiality of things and bodies'.[73] Furthermore, it is an act to remove oneself from his or her community and society, and thus, is often seen to be extremely anti-social and objectionable. This is what Alexander in *The Sacrifice*, the final film of Tarkovsky before he succumbed to terminal lung cancer, is well aware of just before he going to fully commit to it. Awoken from his dream in which he had witnessed the destruction of the world by a full-engaged nuclear war and had taken a vow to sacrifice everything he has to save his family and the world from the annihilation, he sets his house on fire and says frantically to his friend Victor:

> I did it. Don't be upset! Listen to me, Victor. I've got something very impor . . . [*Alexander realises something.*] No! Silence! [*He closes his mouth with his hands.*] Say nothing! Ask nothing![74]

The behaviour that appears to be that of a madman under spell is in fact born of his commitment to not speak. Furthermore, his son, 'Little Man', who was mute due to a throat operation throughout the film, is seen at the end of the film, lying under a dead tree he and Alexander had planted together at the start of the film. Speaking for the first time, he says: 'In the beginning was the Word. Why is that, Papa?'

Compared to these two films that I used as examples of the oral imagery by being mute, the final example of Tarkovsky that I would like to show is rather unsettling. First of all, it is not necessarily about the image of the sealed or the shut mouth, nor is it about being mute; rather it is about gaining a voice. But why is this unsettling? It is unsettling because it shows, I argue, the operation of how the object voice gains power and overtakes the self through language and its signifying chain. In the beginning of *Mirror*, Ignat, the son of Alexei, the protagonist of the film, turns on the TV, showing a documentary on a psychiatric experiment of sorts in which we see a physician (or psychiatrist(?)) and Yuri Zhary, a stutterer in his adolescent years. The short procedure she performs to examine Yuri and cure his impediment is so strange, fascinating, and Foucauldian in its way of incorporating coercion and dominance that I will quote in full:

> Spread your hands. Concentrate. All your tension is centred in your hands. Your hands are strained! Concentrate all of your will power, your big desire to win, on your hands. Your hands are getting more and more tense. They're very tense. Still more tense. Look at your fingers. Your fingers are tense. From here [*her right hand touches the boy's left temple gently*] the tension passes on to your fingers. Look at your hands. Yuri, concentrate! On my count of three your hands will become immobile. One, two, three! Your hands don't move. You can't move them. You're trying to move your hands, but they're fixed. It's very hard for you to make a slightest movement. Now I'm going to lift this transfixion, and you'll be able to speak freely, easily and articulately. [*While her right hand touches the boy's left temple*] From now on you will speak loudly and clearly. Look at me. [*She holds the boy's head gently with both hands.*] I'm lifting the tension from your hands and your speech. One, two, three! [*She pushes the boy's head with force.*] Go ahead, say loudly and clearly: 'I can speak!'[75]

Yuri follows her saying: 'I can speak,' but with much less commitment and certainty than she might have wanted him to do so. Perhaps he does so without stuttering, but the sentence, only three words, is too short for us to confirm it. I cannot forget the final image of Yuri, his troubled eyes looking briefly at the camera just before Tarkovsky abruptly cuts into the main title scene. What is this feeling ill at ease?

Conducting a quick survey on the discourses on stammering or stuttering since antiquity,[76] Connor notes how this speech impediment has been considered a result of either lack or surplus on the tongue. This could be due to an excessive dryness of the tongue, as believed by the Hippocratic school of Kos, or via evasive moisture of the tongue, as illustrated by Galen in medieval Europe. A swelling of the tongue through alcoholic vapours was a suggested cause by Cocles in the fifteenth century, while Francis Bacon gives coldness or dryness, leading to stiffening of the tongue, as the precursor to stammering. Overheating of the tongue was a further reason given by Alexander Ross in the seventeenth century, while breathing through the mouth rather than nose was the sole instigator given by William Abbott in the nineteenth century. At the same time, too much egotism was given as the cause of stammering by Charles Kingsley. Various remedial procedures, either from mechanical theories on the body, or from psychoanalysis, which analysed 'the fantasies invested in the magical omnipotence' and 'fearful failure' of the voice, were provided in order to effect 'normal' speech.[77] Connor argues that in stuttering and other vocal noises, the voice 'meets and mingles with what it is not – indeed is, in the end, nothing more than this mingling'. He continues: '[it is] a fluent mélange in which what it is and what it is not commingle and converse' as the voice 'is nourished by the parasites and imperfections that feed upon it'.[78]

If, following Connor, the voice is an arena in which various vocalic noises that are *outside* the voice are invited into and get mixed into their midst, what is this procedure of eliminating the surplus that is assumed to cause the stammer in the experiment? Has the physician in Tarkovsky's film succeeded in *curing* the impediment? I don't think so. What has happened is that she has caused Yuri to push the pendulum as best as he can so that he is (barely) able to hide those noises (which, based on Connor's survey, look to be categorized as stiffening of the tongue proposed by Bacon). However, we become anxious because we know that once pushed, it may and will pull back, and what is to come is the full force of this inertia – and perhaps this is indeed the point that Tarkovsky intends to highlight.

The lips sealed and the mouth shut are not only symptomatic to, but also evocative of the shutting out of all the modalities of the mouth. As a spiritual, religious or political means, fasting is one such extreme action. In Tibetan Buddhism, a fasting ritual is often accompanied by a vow of silence, both of which is general practice in the universal religious discipline of asceticism. Such a ritual is usually aimed at removing one's worldly concerns and promoting his or her spiritual devotion to life. Jackson points out that while the Buddha objected to 'the life of extreme austerity', the life of monks and nuns is austere, with the mandatory celibacy, fasting and voluntary silence at the centre of their lives. Such fasting rituals, Jackson goes on to argue, were not only beneficial to individuals by purifying negative karma and accumulating merit; they also gave the participants of the rituals a sense of community and belonging.

Fasting, in many Eastern cultural and religious backgrounds, was more a communal than an individual practice:

> On the broadest level, as a collective experience, the fasting ritual provides for the participants a natural sense of community, which ideally will extend beyond the ritual period and find expression in people's ordinary lives. The greater the proportion of a community that participates in the fasting ritual, the greater the ritual's effect on social cohesion will be.[79]

On the other hand, the lips sealed can also bring out a rather drastic and shocking image of lip sewing that has either psychotic or political connotations. It operates within the phantasmagoric modalities of the mouth as it short-circuits the oral imageries to auditory hallucinations, as shown by a prisoner who sewed up his lips and ears because he declared that he had to sew his organs because he was threatened by djinns that his kids would be killed if he were to speak or listen.[80] This practice extends to religious or cultural systems like the Shuar's shrunken heads, also called *tsantsas*, whose lips and eyes had been sewn because of the Shuar's belief that the soul of the killed would come out from the mouth and revenge on them.[81] Similarly, lip sewing was also used during socio-political conflicts witnessed by asylum seekers as a means of protest to force the victim to a death fast.[82] All of these images that the sealed lips and the closed mouth evoke are ultimately connected to the hums of the other, which attempt to relieve the mouth from its overactive, hypersensitive, and voracious modalities. They sever it from its networks; a severance, a sewing-up, a closing-down of the mouth that could ultimately put the self (of the mouth) in danger of living. In this regard, hums are paradoxical and can be thought to be a death drive. Our analysis of the symptoms caused by hums of the other will lead us to the death drive, at which juncture we will show how hums will have us come to face it. But for now, let us take a notion of the death drive proffered by Žižek, who points out that it is 'not a biological fact' as proposed by Freud, but 'a notion indicating that the human psychic apparatus is subordinated to a blind automatism of repetition beyond pleasure-seeking, self-preservation, accordance between man and his milieu. Man is . . . an animal extorted by an insatiable parasite (reason, *logos*, language)'. Thus, he goes on to point out:

> [T]he 'death drive,' this dimension of radical negativity, cannot be reduced to an expression of alienated social conditions, it defines *la condition humaine* as such: there is no solution, no escape from it; the thing to do is not to 'overcome,' to 'abolish' it, but to come to terms with it, to learn to recognize it in its terrifying dimension and then, on the basis of this fundamental recognition, to try to articulate a *modus vivendi* with it.[83]

Hums as symptoms

My fascination with some of the stories from Calvino and Kafka that started this chapter have to do with a realisation that they all somehow highlight a situation where the protagonists become inflicted with hums. Sometimes, these hums appear to them initially as a simple, benign nuisance, and others quickly enrapture them; all hums eventually come to them to pose a threat, which come to question the protagonists' existence in one way or another. With these stories, and the subsequent survey on the hums' mystical and imaginary shades, what I have wanted to show is that these hums are symptoms. These are not impediments that need to be cured, but those that you will have to come to terms with. Not only will they be left uncured, but you may not want them to be completely gone. Designating these signs as symptoms, I do not mean that hums may point to a single (psychiatric, mental or physical) disease or complications that cause them; they are not conditions, contingent upon whether we take care of ourselves or whether we are well or not; rather, the hums are our being-ness, their coming and persisting has nothing to do with our action. In this sense, hums as symptoms are *ontic*,[84] physical, real, existing beings that somehow mysteriously appear in a way that comes simultaneously from us and beyond us. Symptoms of hums are silent, yet addressed to us, prescribing nothing.[85] How peculiar the ways in which hums come to us are! They come as an opening or a blob, a gap or a clot, a silence or a noise, a lack or a surplus. Confronted with them, we always feel their appearance to be strange ('Strange, the same noise there too,' says 'I' in *The Burrow*). But this strangeness, as it turns out, is not necessarily because something new from before has taken place; rather it is more likely that we became, somehow, inexplicably, sensitive to something outside of the familiar. Let us be objective (if we can) and examine the situation 'I' finds himself in in *The Burrow*. What does 'I' really (have to) worry about? It's just a noise, a hum, really. Nothing has happened, the noise has not materialized into anything dangerous. Then why does 'I' contemplate leaving its beloved burrow, which 'I' has taken all of its life to perfect?

In *The Weird and the Eerie*, Fisher argues that the preoccupation with the strange – 'not the horrific' – is what the weird and the eerie have in common. Fisher points out that 'the allure that the weird and the eerie,' which share the quality of being strange, 'is not captured by the idea that we "enjoy what scares us,"' but rather, is 'to do with a fascination for the outside'.[86] If we take Fisher's notion as an answer to the question above (why does 'I' behave the way it does?), we can say that it is not merely things, conditions, situations 'I' is in that causes its behaviour. Rather, it is 'I''s fascination for the outside, which 'I' brought into its psyche and that 'I' became sensitive to, that creates its anxiety. The mixing of the inside and outside, driven by the fascination for the outside, and the boredom of the inside, is a perfect recipe for 'I' in the burrow. In fact, that mixing is almost a principal condition for all protagonists in the stories of Calvino and Kafka that we discussed as well as those of Lovecraft, which Fisher examines in his book (a hero or heroine is bored and discovers or encounters someone – a stranger – or something – a mysterious object – now the plot thickens.) In these stories, the key elements that operate and run all the narrative, no matter what the context may be, are the surplus and the lack, to which the mix of inside and outside makes us become sensitive. Being a surplus, the weird, Fisher points out, is 'that *which does not belong*' and 'brings to the familiar something which ordinarily lies beyond it'. The eerie, however, being a lack, concerns itself with the question of presence and absence, such as '*why is there something here when there should be nothing? Why is there nothing here when there should be something?*' In this way, Fisher continues, both the weird and the eerie 'allow us to see the inside from the perspective of the outside'.[87] Here we are keenly reminded of all the subsequent questions that I ended up identifying from a simple suggestion: *Who am I humming to?* Would it not be that all those questions were, in one way or another, conjugated forms of the same question: why something that does not belong is here, or why something that should belong here is not here? If that is the case, is it not that my initial question of to whom I hum is really an act of mixing the inside and the outside? Then what is inside here and what is the outside?

Here is another symptom of hums that is particularly unique, and yet, quotidian and everyday in the way it is placed in our psyche: *The Hum*. Deming writes that while reports of 'unidentified humming sounds' as a collective experience date back to 1803, it wasn't until the 1970s that so-called 'The Hum' had become more or less recognized as a worldwide phenomenon.[88] Often described as 'a constant throbbing hum, something like the bass frequency of a heavy lorry with its engine on idle', The Hum in twentieth-century reports appears to be quite different sounding than the hums reported in the nineteenth century, such as 'the humming of an apparently large swarm of bees'.[89] The Hum has been experienced by a significant number of people in numerous places in the world, and some of these phenomena[90] have been widely reported by mass media and have brought much interest not only from the general public, but also from scientific communities. Deming points out that despite the name of the Hum, the actual sounds experienced are not a hum, but sounds that are close to 'a diesel engine idling in the distance'. Hence, 'the Hum is not a hum,' he concludes and further qualifies it to be 'a steady hum, a throb, a low speed diesel engine, rumbling and pulsing. A higher pitch [. . .] is sometimes attributed', by quoting Leventhall.[91] Furthermore, Deming offers some potential causes for the Hum: from mass delusion to confounding factors of low-frequency; noises from various machines; symptoms of tinnitus (though a classical symptom of tinnitus is high frequency ringing in the ears); acoustic phenomena, electromagnetic energy turned to sound; cellular phones; the LORAN (Long Range Radio Navigation) system; HAARP (High Frequency Active Auroral Research Program); and TACAMO (Take Charge and Move Out) aircraft.

By briefly introducing Fisher's notion of the weird and the eerie as well as the Hum, a curious and unique symptom of hums of the other, I wanted to highlight that the identical mixing of the inside – hums we make with our mouth and lips closed, hums that are familiar to us, make us calm – and the outside – hums that are not our own making, hums that we have no control of, hums that are completely the other – operates. Along with it, we are left with the same conundrum of that which does not belong, and the same question of presence and absence. The question that many sufferers of the Hum ask in the above case might have been this: *where does this Hum come from?* The lack of the mouth, the sealed lips, the image-voice that conjures up a myriad of oral imaginaries, the ear that has lost control and been given over – all run in their full capacity to conspire and call on us to respond to them: *why is this humming here?*

To answer these key questions of the presence and absence of these hums is effectively and finally to face the essence of hums of the other, to which our discussions have revolved around continuously like an unsayable. By admitting them as symptoms, we are in fact getting close not necessarily to these hums, but to the essence of them. As we have recognized in numerous occasions, hums, as symptoms, persist and keep on humming. While doing so, they mix the inside and the outside; they invite us to hum to ourselves, to our beloved ones, in the shower, in the kitchen, or encourage us to hear hums of others who are close to us, or hubbubs of birds, bees or children playing faraway in the garden. At same time, the symptoms persist and keep returning to the hum, through the refrigerator, through the compressor, cars idling, electricity or water running through the house. The hum may be threatening, preventing us from going to sleep, separating us from others who do not suffer from them, making us vulnerable, hysterical, and weary. Through these hums, we intently and intensively lend our ear – in doing so, it is us, also, who mix the inside and the outside. As we do and as they do, these mixtures bare out the very gap, the lack, the opening, the silence, and the noise, which we do not know how to assimilate into our own. The hums of the other are painfully and utterly on the side of the other; we have no idea how to get to them but they nevertheless come to us, overwhelming our ear smoothly and cunningly, easily and immediately, stealthily and abruptly. How are they here when they should not be here? We suddenly confront hums that are not ours: *have I called on them?* We don't know how to answer it, but then we can always hum it, as if we could hum it away.

One might follow Lacan's advice and take cues from his identification of symptoms, which, according to him, are 'returns of the repressed'.[92] As a psychoanalytic term, symptoms appear initially as traces, which are construed to be effects of a cause made in the past, but their meanings are in fact constructed only with their analysis afterwards. As such, Žižek points out: 'The Lacanian answers to the question "From where does the repressed return?" is therefore, paradoxically, "From the future"'. It is not 'from the hidden depth of the past' that the meanings of symptoms are discovered, but from the analysis, which is 'the signifying frame which gives the symptoms their symbolic place and meaning'.[93] By designating hums, be they from our own mouth or from someone or something else, and even from nowhere, and everywhere, we are admitting, confessing them to be returns of the repressed in Lacanian sense. What is repressed here is of course the Other, the ontological voice of Dasein for Heidegger, the true object voice that is silent, mute. Just as LaBelle admitted that the voice stretches us, hums also stretch us, but this stretching is not our doing. We are being stretched inwards and outwards while our ear continues to be pushed and pulled by hums that travel back and forth between meaning and presence effects; we become nauseous by the speed at and the force with which hums make us run, and yet, we then hum to calm ourselves. This is an existential threat, coming to us without us knowing it has come or that it will have come. The reason this is a threat is not because it will come to us to cause harm. As we have seen, the noise has not, and will never cause harm to 'I' in *The Burrow*, the king in *A King Listens*, Gregor in *The Metamorphosis*, or K. in *The Castle* or indeed numerous people who have heard and suffered from the mysterious *Hum*. The reason that these hums have become a threat is precisely because they come to us, point to the other, a gap, a crack in us, between the subject and his or her voice, which has been bandaged by language or dreams or fantasies. These fantasies work either as a controlling mechanism for the mouth, a looking-away system for our psyche, or a retreat or safe house to place the subject under potential collapse. Ultimately, though, as Žižek has warned us,[94] the terror of hums is less to do with the severance of hums that we can make from our mouth – that is loss of control – or to do with the oral imaginaries and the phantasmagoria of the invisible mouth. Rather, it is to do with the fear of hums *being too close to us and eventually becoming us*. This is the terror that K., under the spell of the buzz, has come to: hums of a thousand children singing from an infinite distance, an impossible blend of echoes running like a single, resonant tone. K. becomes the hum. But this terror of hums come to us less forcefully, starting with the acknowledgement that walls have ears, like the king. The question spawns from 'who else is hearing what I am hearing' to 'what else is hearing', to 'am I hearing that which is hearing', and finally to: *am I only hearing things that which is hearing?* This is the moment of deepest suspicion – our hearing has become completely conditioned and thus controlled by that which is hearing – when I fear that the noise has become me; we have lost our ear, and that which is lending its ear.

How then do we respond to the hums of the other, those that operate exactly like the Lacanian Real, 'conceived as a hard kernel resisting symbolisation, dialecticisation, persisting in its place, always returning to it'?[95] Surprisingly, it is the symptoms themselves that are going to save us from this very terror. Žižek argues:

> What we must bear in mind here is the radical ontological status of symptoms: symptom, conceived as *sinthome*, is literally our only substance, the only positive support of our being, the only point that gives consistency to the subject. In other words, symptom is the way we – the subjects – 'avoid madness', the way we 'choose something (the symptoms-formation) instead of nothing (radical psychotic autism, the destruction of the symbolic universe)' through the binding of our enjoyment to a certain signifying, symbolic formation which assures a minimum of consistency to our being-in-the-world.[96]

Because without hums conceived as symptoms or *sinthomes* in the Lacanian sense, according to Žižek, there would be the other alternative, which 'is nothing: pure autism, a psychic suicide, surrender to the death drive, even to the total destruction of the symbolic universe'.[97] Positioned and posited as such, hums are thought to be a positive force, an affirmative – rather than a negating – manifestation, 'a particular, "pathological", signifying formation, a binding of enjoyment, an inert stain resisting communication and interpretation, a stain which cannot be included in the circuit of discourse, of social bond network, but is at the same time a positive condition of it'.[98] So on the one hand, hums will subsume us, claim us and become us, just like the ones that enraptured K. in *The Castle*; only the difference would be that there won't be a landlord pulling at our coat to break the spell for us; we won't say, 'Go away!' Instead we will be intensely and helplessly listening to them; for how long, we won't know. On the other hand, hums will be what we will come to terms with, just like 'I' in *The Burrow* would have had to come to terms with the noise and the big beast he ended up conjuring, whose arrival he dreaded:

> [I]f I have peace, and danger does not immediately threaten me, I am still quite fit for all sorts of hard work; perhaps, considering the enormous possibilities which its powers of work open before it, the beast has given up the idea of extending its burrow in my direction . . . The more I reflect upon it the more improbable does it seem to me that the beast has even heard me . . . So long as I still knew nothing about it, it simply cannot have heard me, for at that time I kept very quiet, nothing could be more quiet than my return to the burrow; afterwards, when I dug the experimental trenches, perhaps it could have heard me, though my style of digging makes very little noise; but if it had heard me I must have noticed some sign of it, the beast must at least have stopped its work every now and then to listen. But all remained unchanged.[99]

While noting, at the end of his article on Lacan's voice object, the absurdity of how the voice emerges, Miller half-jokingly tells us that the formulation of 'any signifying chain,' would be simply that we say 'to the Other: "Shut up!"'.[100] So perhaps that is what we can do to hums of the other that keep coming to us, uninvited, unannounced, haunting and threatening us; no, we won't yell at them to shut up; remember, our lips are sealed, our mouths shut; the only thing left for us to do is humming; we will keep on humming.

NOTES

Chapter 1

1 Suk-Jun Kim, 'In Tune, Out of Tune', http://berlin.tonspur.at/b_31_en.html (October 2009).

2 You can listen to their hum here at Soundcloud, https://soundcloud.com/suk-jun-kim/silver-city-humming–2012 – from 14:08 minutes onwards.

3 Mark Fisher, *The Weird and the Eerie* (London: Repeater, 2016), 9.

4 'Aberdeen Humming' (November 2013), http://humming-project.donotworkhere.org/.

5 Mladen Dolar, *A Voice and Nothing More (Short Circuits)* (Cambridge: The MIT Press, 2006), 72–3.

6 Ibid., 32.

7 Steven Connor, *Beyond Words: Sobs, Hums, Stutters and Other Vocalizations* (London: Reaktion Books, 2014), 7.

8 Dolar, *A Voice*, 73.

9 Milan Kundera, *The Book of Laughter and Forgetting*, trans. Michael Henry Heim (New York: Penguin, 1981).

10 Listen to David Cameron humming here at BBC: http://www.bbc.co.uk/news/video_and_audio/headlines/36767880/pm-hums-tune-after-may-announcement (last accessed 27 October 2017).

11 Mladen Dolar, 'The Object Voice,' *Gaze and Voice as Love Objects: Sic 1*, ed. Renata Salecl and Slavoj Žižek (Durham: Duke University Press, 1996).

12 Jacques-Alain Miller, 'Jacques Lacan and the Voice', *The Later Lacan: An Introduction (Suny Series in Psychoanalysis and Culture)*, ed. Veronique Voruz (Albany: State University of New York Press, 2007), 137–46.

13 Ibid., 139.

14 Edward S. Casey, *The World At a Glance (Studies in Continental Thought)* (Bloomington: Indiana University Press, 2007), 138.

15 Ibid., 134.

16 The body's role in reflecting the voice to the self is not perfect, though, for that which returns to the self is always the result of an accumulation and subtraction of the voice.

17 Miller, 'Lacan and the Voice', 143.

18 Connor, *Beyond*, 39–40.

19 This is the second time the air sneaks into our discussion of hums earlier that I anticipated, which I take to be the nature of the air!

20 Miller, 'Lacan and the Voice', 139.

21 Dolar, *A Voice*, 73.

22 Ibid., 103.

23 David Novak and Matt Sakakeeny, *Keywords in Sound*, (Durham; London: Duke University Press, 2015), 5.

24 Definition of 'hum', Oxford Dictionaries, https://en.oxforddictionaries.com/definition/hum (accessed 10 January 2018).

25 Miller, 'Lacan and the Voice', 143–4.

26 Ibid., 144.

27 Brandon LaBelle, *Lexicon of the Mouth: Poetics and Politics of Voice and the Oral Imaginary* (London: Bloomsbury Academic, 2014).

28 Brandon LaBelle, *Acoustic Territories: Sound Culture and Everyday Life* (New York: Continuum, 2010), xxv.

29 Slavoj Žižek, *Looking Awry: An Introduction to Jacques Lacan Through Popular Culture (October Books)*, reprint edition (Massachusetts: The MIT Press, 1992), 5–6.

30 Renata Salecl and Slavoj Žižek, 'The Object Voice', *Gaze and Voice as Love Objects (Series: Sic 1)* (Durham: Duke University Press Books, 1996).

31 Alice Lagaay, 'Between Sound and Silence: Voice in the History of Psychoanalysis,' *E-pisteme* 1, no. 1 (2008): 53.

Chapter 2

1 LaBelle, *Lexicon*, 1.

2 Miller, 'Lacan and the Voice', 137–46.

3 LaBelle, *Lexicon*, 5.

4 Ibid., 5.

5 Brouria Bitton-Ashkelony, '"More Interior Than the Lips and the Tongue": John of Apamea and Silent Prayer in Late Antiquity', *Journal of Early Christian Studies* 20, no. 2 (2012): 308.

6 Yael Bentor, 'Consecration of Images and Stûpas in Indo-Tibetan Tantric Buddhism', *Brill's Indological Library* 11 (1997): 88.

7 *The Huntress*, http://deadbydaylight.wikia.com/wiki/The_Huntress (last accessed 12 October 2017).

8 'Dead by Daylight ASMR: Huntress Humming, Rain Fall, Mother Dwelling Sounds', YouTube, https://www.youtube.com/watch?v=0X2F6XQzFkw (accessed 12 October 2017).

9 The Pinocchio Effect posits that 'liars use more words than truth-tellers and that this effect is because liars use more words to control the interaction and provide justification to create a believable reality'; Lyn M. Van Swol, M. T. Braun and D. Malhotra, 'Evidence for the Pinocchio Effect: Linguistic Differences Between Lies, Deception By Omission, and Truth', *Discourse Processes* 49 (2012): 79–106.

10 M. MacSweeney et al., 'Silent Speechreading in the Absence of Scanner Noise: An Event-Related fMRI Study', *Neuroreport* 11 (2000): 1729–33; Sophie Molholm and J.J. Foxe, 'Look "Hear", Primary Auditory Cortex is Active During Lip-Reading', *Neuroreport* 16, no. 2 (2005): 123–4; J. Pekkola et al, 'Primary Auditory Cortex Activation By Visual Speech: an fMRI Study At 3 Tesla', *Neuroreport* 16 (2005): 125–8.

11 J.-L. Schwartz, F. Berthommier and C. Savariaux, 'Audio-Visual Scene Analysis: Evidence for a "Very-Early" Integration Process in Audio-Visual Speech Perception', *Proceedings of the 7th International Conference on Spoken Language Processing* (2002): 1937–40.

12 Sasha Fagel, 'Auditory Speech Illusion Evoked By Moving Lips', *Proceedings of 9th Conference on Speech & Language Processes* (2005): 115–18.

13 Isa. 6.7–8, KJV.

14 1 Cor. 13.1–2, KJV.

15 Bitton-Ashkelony, 'More Interior', 305.

16 Ibid., 307.

17 Ibid., 301.

18 Nigette M. Spikes, *Dictionary of Torture* (Bloomington: Abbott Press, 2015): 121.

19 'Scold's bridle (Germany: 1550–1800)', *Brought to Life: Science Museum*, http://broughttolife.sciencemuseum.org.uk/broughttolife/objects/display?id=5343 (Accessed 25 March 2018).

20 John Cage, *Silence: Lectures and Writings,* MIT Paperback edition (Connecticut: Wesleyan University Press, 1961), 109.

21 Or in 1950; Cage himself was not sure of the exact time of the year, as has been commented in various sources.

22 Paul Sheehan, 'Nothing is More Real: Experiencing Theory in the Texts for Nothing', *Journal of Beckett Studies* 91 (2000): 92.

23 Ibid., 93.

24 Ibid., 92.

25 Ibid., 93.

26 Cage, *Silence*, ix.

27 Richard Brown Jr, 'Lecture on Nothing', *A Year From Monday: Reading Through John Cage's Writings, 2011–2012*, http://www.ayearfrommonday.com/2011/11/lecture-on-nothing-ca-1949-50.html (accessed 10 April 2017).

28 Sheehan, 'Nothing is More Real', 92.

29 Christopher Shultis, *Silencing the Sounded Self: John Cage and the American Experimental Tradition* (Boston: University Press of New England, 2013), 47.

30 Jannika Bock, *Concord in Massachusetts, Discord in the World: The Writings of Henry Thoreau and John Cage (American Culture)*, (Internationaler Verlag der Wissenschaften, 2008), 40.

31 Shultis, *Silencing*, 51.

32 Douglas Kahn, 'John Cage: Silence and Silencing', *The Musical Quarterly* 81, no. 4 (1997): 556–98.

33 Ibid., 581.

34 Ibid., 583.

35 Ibid., 557.

36 Emma Hornby, 'Preliminary Thoughts About Silence in Early Western Chant', *Silence, Music, Silent Music*, ed. Nicky Los and Jenny Doctor (Aldershot: Ashgate, 2007), 151.

37 Ibid., 151.

38 Beth Williamson, 'Sensory Experience in Medieval Devotion: Sound and Vision, Invisibility and Silence', *Spectrum* 88, no. 1 (2013): 31.

39 Ibid., 32.

40 Jeffrey W. Cupchik, 'Buddhism as Performing Art: Visualizing Music in the Tibetan Sacred Ritual Music Liturgies', *Yale Journal of Music and Religion* 1, no. 1 (2015): 32.

41 Ibid., 33.

42 Ibid., 50.

43 Ter Ellingson, 'The Mandala of Sound: Concepts and Sound Structures in Tibetan Ritual Music' (PhD diss., University of Wisconsin-Madison, 1979), 364.

44 Mark Fisher, *Capitalist Realism: Is There No Alternative?* (Winchester: Zero Books, 2009), 18.

45 Renata Salecl and Slavoj Žižek, Object Voice, 93.

46 Cage, *Silence*, 109.

47 Dolar, *A Voice*, 152–7.

48 Ibid., 155.

49 Ibid., 156.

50 Salecl and Žižek, 'Object Voice', 93.

51 Dolar, *A Voice*, 157.

52 Italo Calvino, *Six Memos for the Next Millennium/the Charles Eliot Norton Lectures 1985–86 (Vintage International)* (New York: Vintage Books, 1993), 4.

53 Ibid., 5.

54 Hans Ulrich Gumbrecht, *Production of Presence: What Meaning Cannot Convey* (Stanford: Stanford University Press, 2004), 2.

55 Cage, *Silence*, 128.

56 Gumbrecht, *Production of Presence*, 28–30.

57 Ibid., 30.

58 LaBelle, *Lexicon*.

59 Ibid., 5.

60 Calvino, *Six Memos*, 31–5.

Chapter 3

1 Cage, *Silence*, 108–27.

2 Ibid., 128–45.

3 Italo Calvino, *Invisible Cities* (New York: Harcourt Brace Jovanovich, 1978), 135.

4 I thank Imogene Newland for pointing these out to me.

5 Calvino, *Six Memos*, 5–6.

6 Italo Calvino, 'A King Listens', *Under the Jaguar Sun* (London: Harcourt Brace Jovanovich, 1988), 37–8.

7 Ibid., 44.

8 Ibid., 38.

9 Ibid., 39.

10 Ibid., 38.

11 Franz Kafka, *The Metamorphosis, Franz Kafka, the Complete Stories,* ed. Nahum N. Glatzer (New York: Schocken Books, 1971), 91.

12 Gilles Deleuze and Felix Guattari. *Kafka: Toward a Minor Literature (Theory and History of Literature, 30)* (Minneapolis: University of Minnesota Press, 1986), 13.

13 Gumbrecht, Production of Presence; refer to my discussion on the presence effects and the meaning effects in the previous chapter.

14 Franz Kafka, *The Burrow, Franz Kafka, the Complete Stories,* ed. Nahum N. Glatzer (New York: Schocken Books, 1971), 343.

15 Franz Kafka, *The Castle* (New York: Schocken Books, 1997), 27.

16 Dolar, *A Voice*, 169–170.

17 Miller, 'Lacan and the Voice', 145.

18 LaBelle, *Lexicon*, p. x.

19 Ibid., x.

20 Slavoj Žižek, *The Sublime Object of Ideology* (London: Verso, 1989), 229–34.

21 Ibid., 263.

22 Ibid., 263.

23 David Toop, *Sinister Resonance: The Mediumship of the Listener* (New York: Continuum, 2011), 29.

24 Ibid., 29.

25 Don Ihde, *Listening and Voice: Phenomenologies of Sound*, 2nd ed. (Albany: State University of New York Press, 2007), 94–5.

26 Connor, *Beyond*, 103–4.

27 Ihde, *Listening*, 73.

28 Ibid., 73.

29 Elizabeth Stewart, Maire Jaanus and Richard Feldstein (eds), *Lacan in the German-Speaking World (Suny Series in Psychoanalysis and Culture)* (Albany: State University of New York Press, 2004), 11.

30 Definition of 'aphonic', Oxford Dictionaries, https://en.oxforddictionaries.com/definition/aphonic (accessed 10 February 2018).

31 Definition of 'aphasia', Oxford Dictionaries, https://en.oxforddictionaries.com/definition/aphasia (accessed 10 February 2018).

32 David A. Goode, 'Presentation Practices of a Family With a Deaf-Blind, Retarded Daughter', *Family Relations* 33, no. 1 (1984): 175.

33 Connor, *Beyond*, 10.

34 Toop, *Sinister*, vii–viii.

35 Michel Chion, *Guide Des Objets Sonores: Pierre Schaeffer et la recherche musicale*, trans. John Dack and Christine North, Bibliothèque de recherche musicale (Paris: Institut national de la communication audiovisuelle/Buchet-Chastel, 1983), 18.

36 Brian Kane, *Sound Unseen: Acousmatic Sound in Theory and Practice*, reprint edn (New York: Oxford University Press, 2016), 33–37.

37 Ibid., 7.

38 Suk-Jun Kim, 'A Critique on Pierre Schaeffer's Phenomenological Approaches: Based on the Acousmatic and Reduced Listening', *The International Conference of Pierre Schaeffer: mediArt*, ed. Jerica Ziherl (Rijeka: Museum of Modern and Contemporary Art, 2011), 123–33.

39 Slavoj Žižek, '"I Hear You With My Eyes"; or, the Invisible Master', *Gaze and Voice as Love Objects (Series: Sic 1)*, ed. Renata Salecl and Slavoj Žižek (Durham: Duke University Press Books, 1996) 92.

40 Steven Connor, *Dumbstruck: A Cultural History of Ventriloquism* (Oxford: Oxford University Press, 2000).

41 Ibid., 35.

42 Dolar, *A Voice*, 60.

43 Ibid., 68–9.

44 Miller, 'Lacan and the Voice', 137–46.

45 Dolar, *A Voice*, 161.

46 Ibid., 161.

47 Kane, *Sound*, 202.

48 Ibid., 203.

49 Definition of 'hum', Oxford Dictionaries, https://en.oxforddictionaries.com/definition/hum (accessed 10 January 2018).

50 Steven Connor, *The Matter of Air: Science and Art of the Ethereal* (London: Reaktion Books, 2010), 10.

51 Ibid., 178–9.

52 Ibid., 179.

53 Christine L. Corton, *London Fog: The Biography* (Cambridge: Harvard University Press, 2015).

54 Connor, *Matter*, 179.

55 Ibid., 65.

56 Luce Irigaray, *The Forgetting of Air in Martin Heidegger*, trans. Mary Beth Mader (London: The Athlone Press, 1999), 6.

57 Connor, *Beyond*, 125.

58 Connor, *Matter*, 15.

59 Ibid., 15.

60 Ibid., 16.

61 Ibid., 30–1.

62 Ibid., 35.

63 Irigaray, *Forgetting*, 8.

64 Ibid., 9.

65 Gaston Bachelard, *Air and Dreams: An Essay on the Imagination of Movement (Bachelard Translation Series)* (Dallas Institute Publications, 2011): 13.

66 Ibid., 21.

67 Connor, *Matter*, 35.

68 Connor, *Beyond*, 104.

69 Dolar, *A Voice*, 152.

70 Michael A. Sells, *Mystical Languages of Unsaying*, 1st edn (Chicago: University of Chicago Press, 1994) 2–3.

71 *Saint Augustine: On Christian Doctrine*, trans. D.W. Robertson Jr. (Indianapolis: The Bobbs Merrill Company, 1958), 10–11.

72 *Andrei Rublev*, directed by A. Tarkovsky (USSR: Mosfilm Studio, 1966), DVD.

73 LaBelle, *Lexicon*, 2.

74 *The Sacrifice*, directed by A. Tarkovsky (Sweden/France: The Swedish Film Institute/Argos Films, 1986), DVD.

75 *Mirror*, directed by A. Tarkovsky (USSR: A Mosfilm Unit 4 Production, 1974), DVD.

76 LaBelle, *Lexicon*, 18–32.

77 Connor, *Beyond*, 22.

78 LaBelle, *Lexicon*, 32.

79 Roger Jackson, 'A Fasting Ritual', *Religions of Tibet in Practice*, ed. Donald S. Lopez Jr. (Princeton: Princeton University Press, 1997), 275.

80 S. Taktak, S. Ersoy, A. Ünsal and M. Yetkiner, 'The Man Who Sewed His Mouth and Ears: A Case Report', *Health Care: Current Reviews* 2, no. 2 (2014).

81 Steven Lee Rubenstein, 'Circulation, Accumulation, and the Power of Shuar Shrunken Heads', *Cultural Anthropology* 22, no. 3 (2007): 357–99; 'The Amazon's head hunters and body shrinkers', *The Week*, 20 January 2012, http://theweek.com/articles/478804/amazons-head-hunters-body-shrinkers (accessed 10 February 2018).

82 Emma Cox, *Performing Noncitizenship: Asylum Seekers in Australian Theatre, Film and Activism (Anthem Australian Humanities Research Series)* (New York: Anthem Press, 2015), 117–23. Also refer to Catherine Adams and Tania Branigan, 'Refugee sews up his lips, eyes and ears', *Guardian*, 27 May 2003, https://www.theguardian.com/uk/2003/may/27/immigration.immigrationpolicy; 'Prostitutes sew lips together in Bolivia protest', *Reuters*, 25 October 2007, https://www.reuters.com/article/us-bolivia-prostitutes/prostitutes-sew-lips-together-in-bolivia-protest-idUSN2436073120071024; 'Calais migrants sew lips shut in protest against destruction of camp', *The Journal*, 4 March 2016, http://www.thejournal.ie/calais-lips-2641887-Mar2016/ (accessed 10 February 2018).

83 Žižek, *Sublime*, xxvii–xxviii.

84 'Relating to entities and the facts about them; relating to real as opposed to phenomenal existence', https://en.oxforddictionaries.com/definition/ontic (accessed 20 February 2018).

85 Kane, *Sound*, 205.

86 Mark Fisher, *The Weird and the Eerie* (London: Repeater, 2016), 8.

87 Ibid., 10–12.

88 'The World Hum Map and Database Project', http://www.thehum.info/ (accessed 10 July 2017).

89 D. Deming, 'The Hum: An Anomalous Sound Heard Around the World', *Journal of Scientific Exploration* 18, no. 4 (2004): 571–95; quoted from W. R. Corliss, 'Earthquakes, Tides, Unidentified Sounds and Related Phenomena', *A Catalogue of Geophysical Anomalies* (Glen Arm: The Sourcebook Project, 1983) 179.

90 In his article, Deming surveys these places and related reports, including the Bristol Hum, the Largs Hum, the Taos Hum, the Kokomo Hum, and so on.

91 Deming, 'The Hum', 575; G. Leventhall, *A Review of Published Research on Low Frequency Noise and Its Effects* (London: Department for Environment, Food and Rural Affairs, 2003), 43.

92 Žižek, *Sublime*, 57.

93 Ibid., 58.

94 Žižek, 'I Hear You', 92–3.

95 Žižek, *Sublime*, 181.

96 Ibid., 81.

97 Ibid., 81.

98 Ibid., 81–2.

99 Kafka, *Burrow*, 359.

100 Miller, 'Lacan and the Voice', 145.

BIBLIOGRAPHY

Bachelard, Gaston. *Air and Dreams: An Essay on the Imagination of Movement*. Bachelard Translation Series. Dallas: Dallas Institute Publications, 2011.

Bentor, Yael. 'Consecration of Images and Stûpas in Indo-Tibetan Tantric Buddhism'. *Brill's Indological Library* 11. Leiden: Brill, 1997.

Bitton-Ashkelony, Brouria. '"More Interior Than the Lips and the Tongue": John of Apamea and Silent Prayer in Late Antiquity.' *Journal of Early Christian Studies* 20, no. 2 (Summer 2012): 303–31.

Bock, Jannika. *Concord in Massachusetts, Discord in the World: The Writings of Henry Thoreau and John Cage (American Culture)*. 1st edn. Peter Lang GmbH, Internationaler Verlag de Wissenschaften, 2008.

Brown, Richard Jr. 'Lecture on Nothing.' *A Year From Monday: Reading Through John Cage's Writings, 2011–2012*. Accessed 10 April 2017. http://www.ayearfrommonday.com/2011/11/lecture-on-nothing-ca-1949-50.html.

Cage, John. *Silence Lectures and Writings*. First MIT paperback edition. Middletown: Wesleyan University Press, 1961.

Calvino, Italo. *Invisible Cities*. 1st Harvest/HBJ edn. New York: Harcourt Brace Jovanovich, 1978.

Calvino, Italo. 'A King Listens'. *Under the Jaguar Sun*. London: Harcourt Brace Jovanovich, 1988.

Calvino, Italo. *Six Memos for the Next Millennium/the Charles Eliot Norton Lectures 1985–86 (Vintage International)*. New York: Vintage, 1993.

Casey, Edward S. *The World At a Glance (Studies in Continental Thought)*. Bloomington: Indiana University Press, 2007.

Connor, Steven. *Dumbstruck: A Cultural History of Ventriloquism*. Oxford: Oxford University Press, 2000.

Connor, Steven. *The Matter of Air: Science and Art of the Ethereal*. London: Reaktion Books, 2010.

Connor, Steven. *Beyond Words: Sobs, Hums, Stutters and Other Vocalizations*. London: Reaktion Books, 2014.

Corliss, William R. 'Earthquakes, Tides, Unidentified Sounds and Related Phenomena'. *A Catalog of Geophysical Anomalies*. Glen Arm: The Sourcebook Project, 1983.

Corton, Christine L. *London Fog: The Biography.* Cambridge: Harvard University Press, 2015.
Cox, Emma. *Performing Noncitizenship: Asylum Seekers in Australian Theatre, Film and Activism (Anthem Australian Humanities Research Series).* New York: Anthem Press, 2015.
Cupchik, Jeffrey W. 'Buddhism as Performing Art: Visualizing Music in the Tibetan Sacred Ritual Music Liturgies.' *Yale Journal of Music and Religion* 1, no. 1 (2015).
Deleuze, Gilles, and Felix Guattari. *Kafka: Toward a Minor Literature. Vol. 30, Theory and History of Literature.* Minneapolis: University of Minnesota Press, 1986.
Deming, David. 'The Hum: An Anomalous Sound Heard Around the World.' *Journal of Scientific Exploration* 18, no. 4 (2004): 571–95.
Dolar, Mladen. 'The Object Voice.' *Gaze and Voice as Love Objects: Sic 1.* Edited by Renata Salecl and Slavoj Žižek. Durham: Duke University Press Books, 1996.
Dolar, Mladen. *A Voice and Nothing More (Short Circuits).* Cambridge: The MIT Press, 2006.
Ellingson, Ter. 'The Mandala of Sound: Concepts and Sound Structures in Tibetan Ritual Music'. PhD diss., University of Wisconsin-Madison, 1979.
Fagel, Sascha. 'Auditory Speech Illusion Evoked By Moving Lips'. *Proceedings of 9th Conference on Speech & Language Processes* (2005): 115–18.
Fisher, Mark. *Capitalist Realism: Is There No Alternative?* Winchester: Zero Books, 2009.
Fisher, Mark. *The Weird and the Eerie.* London: Repeater, 2016.
Goode, David A. 'Presentation Practices of a Family With a Deaf-Blind, Retarded Daughter'. *Family Relations* 33, no. 1 (1984): 173–85.
Gumbrecht, Hans Ulrich. *Production of Presence: What Meaning Cannot Convey.* Stanford: Stanford University Press, 2004.
Hornby, Emma. 'Preliminary Thoughts About Silence in Early Western Chant'. *Silence, Music, Silent Music.* Edited by Nicky Los and Jenny Doctor. Aldershot: Ashgate, 2007.
Ihde, Don. *Listening and Voice: Phenomenologies of Sound.* 2nd ed. Albany: State University of New York Press, 2007.
Irigaray, Luce. *The Forgetting of Air in Martin Heidegger.* Translated by Mary Beth Mader. London: The Athlone Press, 1999.
Jackson, Roger. 'A Fasting Ritual'. *Religions of Tibet in Practice.* Edited by Donald S. Lopez Jr. Princeton: Princeton University Press, 1997. 271–92.
Kafka, Franz. *The Burrow. Franz Kafka, the Complete Stories.* Edited by Nahum N. Glatzer. New York: Schocken Books, 1971.
Kafka, Franz. *The Metamorphosis. Franz Kafka, the Complete Stories.* Edited by Nahum N. Glatzer New York: Schocken Books, 1971.

Kafka, Franz. *The Castle*. New York: Schocken Books, 1997.
Kahn, Douglas. 'John Cage: Silence and Silencing'. *The Musical Quarterly* 81, no. 4 (1997): 556–98.
Kane, Brian. *Sound Unseen: Acousmatic Sound in Theory and Practice*. Reprint edn. New York: Oxford University Press, 2016.
Kim, Suk-Jun. 'A Critique on Pierre Schaeffer's Phenomenological Approaches: Based on the Acousmatic and Reduced Listening'. *The International Conference of Pierre Schaeffer: mediArt*. Edited by Jerica Ziherl. 1 Rijeka: Museum of Modern and Contemporary Art, 2011. 123–33.
Kundera, Milan. *The Book of Laughter and Forgetting*. Translated by Michael Henry Heim. New York: Penguin, 1981.
LaBelle, Brandon. *Acoustic Territories: Sound Culture and Everyday Life*. New York: Continuum, 2010.
LaBelle, Brandon. *Lexicon of the Mouth: Poetics and Politics of Voice and the Oral Imaginary.* London: Bloomsbury Academic, 2014.
Lagaay, Alice. 'Between Sound and Silence: Voice in the History of Psychoanalysis'. *E-pisteme* 1, no. 1 (2008): 53–62.
Leventhall, Geoff. *A Review of Published Research on Low Frequency Noise and Its Effects*. London: Department for Environment, Food and Rural Affairs, 2003.
MacSweeney, M., E. Amaro, G.A. Calvert, R. Campbell, A.S. David, P. McGuire, S.C. Williams et al. 'Silent Speechreading in the Absence of Scanner Noise: An Event-Related fMRI Study'. *Neuroreport* 11 (2000): 1729–33.
Miller, Jacques-Alain. 'Jacques Lacan and the Voice'. *The Later Lacan: An Introduction (Suny Series in Psychoanalysis and Culture)*. Edited by Veronique Voruz. 1st edition. Albany: State University of New York Press, 2007. 137–46.
Molholm, Sophie and John J. Foxe. 'Look "hear", Primary Auditory Cortex is Active During Lip-Reading'. *Neuroreport* 16 (2005): 123–24.
Novak, David and Matt Sakakeeny. *Keywords in Sound*. Durham; London: Duke University Press, 2015.
Pekkola, J., V. Ojanen, T. Autti, I.P. Jääskeläinen, R. Möttönen, A. Tarkiainen and M. Sams. 'Primary Auditory Cortex Activation By Visual Speech: An fMRI Study At 3 Tesla', *Neuroreport* 16 (2005): 125–28.
Robertson, D.W., trans. *Saint Augustine: On Christian Doctrine*. Indianapolis: The Bobbs Merrill Company, 1958.
Rubenstein, Steven Lee. 'Circulation, Accumulation, and the Power of Shuar Shrunken Heads', *Cultural Anthropology* 22, no. 3 (2007): 357–99.
Schwartz, J.-L., F. Berthommier and C. Savariaux. 'Audio-Visual Scene Analysis: Evidence for a "Very-Early" Integration Process in

Audio-Visual Speech Perception'. *Proceedings of the 7th International Conference on Spoken Language Processing* (2002): 1937–40.
Sells, Michael A. *Mystical Languages of Unsaying*. 1st edn. Chicago: University of Chicago Press, 1994.
Sheehan, Paul. 'Nothing is More Real: Experiencing Theory in the Texts for Nothing'. *Journal of Beckett Studies* 91 (2000): 89–104.
Shultis, Christopher. *Silencing the Sounded Self: John Cage and the American Experimental Tradition*. Boston: University Press of New England, 2013.
Spikes, Nigette M. *Dictionary of Torture*. Bloomington: Abbott Press, 2015.
Stewart, Elizabeth, Maire Jaanus and Richard Feldstein. *Lacan in the German-Speaking World (Suny Series in Psychoanalysis and Culture)*. 1st edn. Albany: State University of New York Press, 2004.
Taktak, S., S. Ersoy, A. Ünsal, and M. Yetkiner. 'The Man Who Sewed His Mouth and Ears: A Case Report'. *Health Care: Current Reviews* 2, no. 2 (2014).
Toop, David. *Sinister Resonance: The Mediumship of the Listener*. New York: Continuum, 2011.
Van Swol, L. M., M. T. Braun and D. Malhotra. 'Evidence for the Pinocchio Effect: Linguistic Differences Between Lies, Deception By Omission, and Truth'. *Discourse Processes* 49 (2012): 79–106.
Williamson, Beth. 'Sensory Experience in Medieval Devotion: Sound and Vision, Invisibility and Silence'. *Spectrum* 88, no. 1 (2013).
Žižek, Slavoj. *The Sublime Object of Ideology (The Essential Žižek)*. Verso, 1989.
Žižek, Slavoj. *Looking Awry: An Introduction to Jacques Lacan Through Popular Culture (October Books)*. Reprint edn. Cambridge, MA: The MIT Press, 1992.
Žižek, Slavoj. '"I Hear You With My Eyes"; or, the Invisible Master'. *Gaze and Voice as Love Objects: Sic 1*. Edited by Renata Salecl and Slavoj Žižek. Durham: Duke University Press Books, 1996. 90–126.

Film

Tarkovsky, Andrei, dir. *Andrei Rublev*. 1966; USSR: Mosfilm Studio. DVD.
Tarkovsky, Andrei, dir. *Mirror*. 1974; USSR: A Mosfilm Unit 4 Production. DVD.
Tarkovsky, Andrei, dir. *The Sacrifice*. 1986; Sweden/France: The Swedish Film Institute/Argos Films in association with Film Four International. DVD.

INDEX